Weighted polynomial approximation methods for Cauchy singular integral equations in the non-periodic case

von der Fakultät für Mathematik
der Technischen Universität Chemnitz genehmigte

Dissertation

zur Erlangung des akademischen Grades
Doctor rerum naturalium
(Dr. rer. nat.)

vorgelegt von

Dipl.-Math. Uwe Weber

geboren am 15. Juni 1968 in Karl-Marx-Stadt (jetzt Chemnitz)

eingereicht am 19. 7. 1999

Gutachter: Prof. Dr. Peter Junghanns
Prof. Dr. David Elliott
Dr. Steffen Roch

Tag der Verteidigung: 20. 12. 1999

Die Deutsche Bibliothek - CIP-Einheitsaufnahme

Weber, Uwe:
Weighted polynomial approximation methods for Cauchy singular integral equations in the nonperiodic case / vorgelegt von Uwe Weber. - Berlin : Logos-Verl., 2000
Zugl.: Chemnitz, Techn. Univ., Diss., 1999
ISBN 3-89722-352-X

ISBN 3-89722-352-X

Logos Verlag Berlin
Michaelkirchstr. 13
10179 Berlin
Tel.: 030 - 42851090
INTERNET: http://www.logos-verlag.de

Bibliographische Beschreibung

Weber, Uwe

Weighted polynomial approximation methods for Cauchy singular integral equations in the non-periodic case

Dissertation, 115 Seiten, TU Chemnitz, Fakultät für Mathematik, 1999.

Abstract

In the present paper a new approach to the numerical solution of Cauchy singular integral equations on the interval by collocation and Galerkin methods is considered. Both methods are based on weighted orthogonal polynomials. The main advantage of our approach, in particular of the collocation method, is the fact that in contrast to usual methods its construction does not depend on the concrete equation and requires less preprocessing. Furthermore, it can also be applied to the system case. On the basis of Banach algebra methods, necessary and sufficient stability conditions are derived, where from the coefficients of the operator only piecewise continuity is required. In a scale of Sobolev spaces we can prove results on convergence rates. Furthermore, in the case of the collocation method we discuss some computational aspects to derive effective algorithms for the fast solution of the approximate equations and present numerical results. In the case of equations perturbed by an integral operator with smooth kernel, we consider a quadrature method and two fast algorithms that allow to make use of the fast solution of the unperturbed equation by the collocation method.

Keywords

Banach algebra methods, Cauchy singular integral equation, Collocation method, Fast algorithm, Galerkin method, Local principle, Orthogonal polynomials, Quadrature method, Sobolev space, Stability.

Acknowledgements

An erster Stelle möchte ich meinem Betreuer, Prof. Peter Junghanns, für sein ständiges Interesse am Entstehen der vorliegenden Arbeit, seine zahlreichen wertvollen Ratschläge und seine moralische Unterstützung herzlich danken.

Weiterhin gebührt mein Dank allen Kollegen, die durch nützliche Hinweise und Diskussionen diese Arbeit unterstützt haben, darunter Frau Dr. Maria Rosaria Capobianco, Herrn Dr. Hans-Jürgen Fischer, Frau Dr. Karla Rost und insbesondere meinem Freund Dr. Uwe Luther.

Ich danke weiterhin Herrn Prof. Viktor Didenko, Herrn Prof. David Elliott, Herrn Dr. Michael A. Golberg und Herrn Dr. Steffen Roch dafür, daß sie sich als Gutachter für diese Arbeit zur Verfügung gestellt haben.

Schließlich danke ich auch meinen Eltern für ihre vielfältige Unterstützung.

Contents

1 Introduction

The subject of the present paper is the investigation of numerical methods based on weighted polynomials for the approximate solution of singular integral equations on $(-1,1)$ of the type

$$(Au)(x) = a(x)u(x) + \frac{b(x)}{\pi i}\int_{-1}^{1}\frac{u(t)}{t-x}\,dt = f(x), \qquad x \in (-1,1), \tag{1.1}$$

where u is the unknown function and a, b, f are given. The integral is to be understood in the sense of the Cauchy principal value. All functions involved are assumed to be complex-valued.

Two sections are also devoted to the more general equation

$$(Au)(x) + \int_{-1}^{1} k(t,x)u(t)\,dt = f(x), \qquad x \in (-1,1), \tag{1.2}$$

where A is as in (1.1). Equations of these types occur in several fields of physics and engineering, such as aerodynamics, hydrodynamics, and elasticity theory (cf. [41], [26], [34]).

Let us introduce some notations that will be used in the following. By $\mathbb{R}$ and $\mathbb{C}$ we denote the real and the complex number field, respectively, and $\mathbb{T}$ is the unit circle line $\{t \in \mathbb{C} : |t| = 1\}$. Let $\Gamma \subset \mathbb{C}$ be a Lyapunov curve. By $\mathbf{L}^\infty(\Gamma)$ and $\mathbf{L}^p(\Gamma)$ $(p \geq 1)$ we denote the usual Lebesgue spaces of all (classes of) measurable complex-valued functions that are essentially bounded or the p-th power of which is Lebesgue integrable, respectively, equipped with the norms

$$\|f\|_\infty = \operatorname{ess\,sup}_{t\in\Gamma} |f(t)|,$$

$$\|f\|_{\mathbf{L}^p(\Gamma)} = \left(\int_\Gamma |f(t)|^p\,dt\right)^{\frac{1}{p}}.$$

The Cauchy singular integral operator on Γ is defined by the equation

$$(S_\Gamma u)(x) = \frac{1}{\pi i}\int_\Gamma \frac{u(t)}{t-x}\,dt, \tag{1.3}$$

where the integral is to be understood in the sense of the principal value. It is well-known ([20], Theorems I.2.1 and I.3.1) that S_Γ can be extended to a bounded linear operator on $\mathbf{L}^2(\Gamma)$. As usual, we also introduce the operators

$$P_\Gamma = \frac{1}{2}(I + S_\Gamma), \qquad Q_\Gamma = \frac{1}{2}(I - S_\Gamma).$$

In the case of a closed curve Γ, the operators P_Γ, Q_Γ are complementary projections ([20], Th. I.1.1) . If Γ is the interval $[-1,1]$, we will in the following often omit the index referring to Γ.

Equation (1.1) will be considered in the weighted Lebesgue space $\mathbf{L}^2_\sigma := \mathbf{L}^2_\sigma(-1,1)$ of all (classes of) measurable functions $u : (-1,1) \to \mathbb{C}$ for which

$$\|u\|_\sigma^2 := \int_{-1}^{1} |u(x)|^2 \sigma(x)\,dx$$

is finite, equipped with the inner product

$$\langle u, v\rangle_\sigma := \int_{-1}^{1} u(x)\overline{v(x)}\sigma(x)\,\mathrm{d}x,$$

which turns $\mathbf{L}^2_\sigma$ into a Hilbert space. Here σ is a Jacobi weight defined by $\sigma(x) = v^{\alpha,\beta}(x) := (1-x)^\alpha(1+x)^\beta$ and satisfying the conditions

$$-1 < \alpha, \beta < 1. \tag{1.4}$$

The conditions (1.4) guarantee the boundedness of the operator S on $\mathbf{L}^2_\sigma$ ([20], Theorem I.4.1). The coefficients a, b are assumed to belong to the set $\mathbf{PC}[-1,1]$ of all piecewise continuous functions. Here $\mathbf{PC}(\Gamma)$ is defined as the closure (in the space of all bounded complex-valued functions, equipped with the supremum norm) of the set of those functions being continuous on Γ with the possible exception of a finite number of jumps, where it is assumed that the value of the function in every point coincides with one of the one-sided limits as the argument approaches this point. (Note that $\mathbf{PC}$-functions possess finite one-sided limits at all points.) For the sake of definiteness we agree for $f \in \mathbf{PC}[-1,1]$ that $f(x_0) = \lim\limits_{x\to x_0-0} f(x)$ for $-1 < x_0 \le 1$ and f is continuous in $x=-1$. For functions in $\mathbf{PC}(\mathbb{T})$ it will turn out to be convenient (cf. Section 5) if we specify

$$f(e^{is_0}) = \lim_{s\to s_0+0} f(e^{is}), \qquad 0 \le s_0 < \pi,$$

$$f(e^{is_0}) = \lim_{s\to s_0-0} f(e^{is}), \qquad \pi \le s_0 < 2\pi.$$

Under these assumptions, the operator on the left-hand side of (1.1), which in the following will be briefly referred to as $aI + bS$, is bounded on $\mathbf{L}^2_\sigma$.

At this point we recall the definition of the spaces $\mathbf{C}^{m,\mu}(\Gamma)$ (for an infinitely smooth curve Γ), where $m \ge 0$ is an integer and $0 \le \mu \le 1$. They consist of all functions that are m times continuously differentiable, where the m-th derivative satisfies a Hölder condition with exponent μ. As usual, the norm in this space is given by

$$\|g\|_{\mathbf{C}^{m,\mu}} = \sum_{j=0}^{m} \|g^{(j)}\|_\infty + \sup_{x\ne y} \frac{|g^{(m)}(x) - g^{(m)}(y)|}{|x-y|^\mu}.$$

if $\mu > 0$. If $\mu = 0$, the second summand in this definition is omitted. In this case we will use the notations $\mathbf{C}^m := \mathbf{C}^{m,0}$, $\mathbf{C} := \mathbf{C}^0$.

Proposition 1.1 ([44], Prop. 9.7, Theorem 9.9)
Assume that $\widetilde{a}, \widetilde{b} \in \mathbf{C}^{0,\eta}[-1,1]$, $0 < \eta < 1$, are real-valued functions for which

$$\widetilde{a}(x) - i\widetilde{b}(x) \ne 0 \qquad \textit{for all} \quad x \in [-1,1].$$

Let λ, ν be integers satisfying the relations

$$\alpha_0 := \lambda + g(1), \quad \beta_0 := \nu - g(-1) \in (-1,1),$$

where g is a continuous function such that

$$\widetilde{a}(x) - i\widetilde{b}(x) = \sqrt{\widetilde{a}^2(x) + \widetilde{b}^2(x)}\, e^{i\pi g(x)}.$$

Then there exists a positive function $w \in \mathbf{C}^{0,\eta}$ such that the singular integral operator $(\widetilde{a}I + iS\widetilde{b}I)v^{\alpha_0,\beta_0}wI$ transforms every polynomial of degree n into a polynomial of degree $n-\kappa$, where $\kappa = -(\lambda+\nu)$. (If $n < \kappa$, this is to be understood in the sense that a polynomial of negative degree is identically zero.)

A lot of papers has been devoted to the numerical solution of equations of the types (1.1) and (1.2) by means of Galerkin, collocation, and quadrature methods, for instance [10], [13], [12], [30], [44, Chapter 9]. The usual approach considered in these papers is based on mapping properties as in Proposition 1.1. An essential drawback of these methods is the fact that in general one has to determine the parameters of Gaussian quadrature formulas associated with generalized Jacobi weights that are related to the coefficients a and b of the operator. This in turn requires the computation of the recurrence coefficients of the orthogonal polynomials with respect to such weights, where the Stieltjes algorithm (cf. [17], [26]) can be applied. Thus, one has considerable computational complexity in the preprocessing. The numerical methods proposed in our paper are independent of the concrete equation and are based exclusively on pure Jacobi weights, for which there exist explicit (and very simple) expressions for the recurrence coefficients. If one uses the Chebyshev weights of first to fourth kind (that is, $v^{\alpha,\beta}$ with $|\alpha| = |\beta| = \frac{1}{2}$) even the weights and nodes of the associated quadrature formulas are known. In the paper [14], Chebyshev polynomials are used to represent the approximate solution, but the problem of determining parameters associated with generalized Jacobi polynomials still remains present.

A second disadvantage of the traditional methods is the fact that the mapping properties that they are based on are dependent on a certain smoothness of the coefficients a and b. When considering, for example, certain free boundary value problems for seepage in channels, one is lead to nonlinear Cauchy singular integral equations that are treated numerically by applying a collocation method and solving the nonlinear discrete equations obtained in this way by a Newton iteration method (see, for example, [29], [24], [26], [27]). This can also be interpreted as the solution of a sequence of linear equations similar to (1.1) by a collocation method, and in the practical applications it is conceivable that the coefficients of these equations have discontinuities. This was a second motivation for considering our method, since it requires only piecewise continuous coefficients.

Last but not least, in contrast to methods based on invariance relations like Proposition 1.1, which work only in the scalar case, the method presented here is also applicable to systems.

Our approach will be based on Banach algebra methods that make use of the fact that the stability of an operator sequence can be formulated equivalently as an invertibility problem in a Banach algebra. The investigation of approximation methods by Banach algebra methods has a long tradition and is a powerful tool in particular when considering equations with discontinuous coefficients. First, in [35] a Galerkin method for multidimensional discrete convolution equations with continuous symbol was investigated. The paper [49] introduced the idea of essentialization (that is, lifting theorems) and treated a Galerkin method for Toeplitz operators with piecewise continuous symbol. In [50] and [31] the Galerkin method and the collocation method for singular integral equations on the unit circle were considered. The present paper is now devoted to transferring these methods to the non-periodic case and combining them with the results achieved with respect to the polynomial approximation methods for this case mentioned above.

The paper is organized as follows: After the description of the approximation methods we have in mind in Section 2 we summarize some facts on Banach algebra methods in Section 3. The following part is devoted to the collocation method: We show the strong convergence of the operator sequences involved in Section 4, derive the main stability result in Section 5 and briefly treat the system case in Section 6. In Section 7 a scale of Sobolev spaces, in which we will give error estimates, is introduced. Then we discuss some computational aspects and present numerical results. The following two sections are devoted to the quadrature method and two fast algorithms for the solution of (1.2). Then we treat the Galerkin method, and in the last section we finally mention some open problems. At the end of the paper we give an index of the symbols and notations.

2 Some further notations and preliminaries

Let v denote a further Jacobi weight satisfying (1.4). (We will begin with considering this general situation, whereas later on the consideration of some aspects will be restricted to special cases of σ and v. The reference to the weight v will be omitted in most of the notations.) By p_n^v we denote the orthonormal polynomial of degree n (with positive leading coefficient) with respect to the inner product $\langle .,. \rangle_v$, that is, $\langle p_n^v, p_m^v \rangle_v = \delta_{mn}$, where δ_{mn} is the Kronecker symbol. Let us consider at this place two special systems of orthogonal polynomials that will play a particular role in our considerations and will therefore be given special notations. Denote by φ the Chebyshev weight of second kind

$$\varphi(x) = \sqrt{1-x^2}$$

and introduce the orthonormal polynomials $U_n := p_n^{\varphi}$ and $T_n := p_n^{\varphi^{-1}}$. For these polynomials we have the well-known trigonometric representations

$$U_n(\cos s) = \sqrt{\frac{2}{\pi}} \frac{\sin(n+1)s}{\sin s}, \qquad n = 0, 1, \ldots$$

$$T_0 = \sqrt{\frac{1}{\pi}}, \qquad T_n(\cos s) = \sqrt{\frac{2}{\pi}} \cos ns, \qquad n = 1, 2, \ldots . \tag{2.1}$$

With respect to the Chebyshev polynomials, the following mapping properties of weighted singular integral operators hold, which are a special case of Proposition 1.1 ([44], 9.15)

$$S\varphi U_n = iT_{n+1}, \qquad n = 0, 1, \ldots,$$

$$S\varphi^{-1}T_0 = 0, \qquad S\varphi^{-1}T_n = -iU_{n-1}, \qquad n = 1, 2, \ldots . \tag{2.2}$$

We introduce the function

$$w_{v,\sigma^{-1}} := \sqrt{\sigma^{-1}v}.$$

Obviously, $w_{v,\sigma^{-1}}I$ is an isometric isomorphism from $\mathbf{L}_v^2$ onto $\mathbf{L}_\sigma^2$. Thus, the functions

$$\widetilde{u}_n := w_{v,\sigma^{-1}} p_n^v, \qquad n = 0, 1, 2, \ldots, \tag{2.3}$$

form an orthonormal basis in $\mathbf{L}_\sigma^2$, because the same is true for p_n^v in the space $\mathbf{L}_v^2$. For the approximation methods we want to apply to (1.1), the functions (2.3) will be used as ansatz functions. We need two sequences of projections with respect to the system $\{\widetilde{u}_n\}_{n=0}^{\infty}$. The first are the Fourier projections P_n^σ given by

$$P_n^\sigma u := \sum_{k=0}^{n-1} \langle u, \widetilde{u}_k \rangle_\sigma \widetilde{u}_k.$$

Evidently, P_n^σ converges strongly to the identity operator I as $n \to \infty$. The second is the weighted interpolation operator $\widetilde{L_n^\sigma}$ assigning to a Riemann integrable function f the uniquely determined weighted polynomial of degree less than n (that is, the element of $X_n := \operatorname{span}\{\widetilde{u}_k\}_{k=0}^{n-1}$) that coincides with f in the collocation points x_{jn}^v $(j = 1, \dots, n)$, which are the zeros of p_n^v. Thus we can write

$$\widetilde{L_n^\sigma} = w_{v,\sigma^{-1}} L_n^v (w_{v,\sigma^{-1}})^{-1} I,$$

where L_n^v is the usual (polynomial) Lagrangian interpolation operator with respect to these nodes. A class of functions f for which $\|\widetilde{L_n^\sigma} f - f\|_\sigma \to 0$ will be described in Corollary 4.5.

We now consider two types of numerical methods to solve equation (1.1) approximately. In both cases an approximate solution $u_n \in X_n$ is sought. The first one is a Galerkin method (also called finite section method), where (1.1) is replaced by the approximate equations

$$\langle f - Au_n, \widetilde{u}_k \rangle_\sigma = 0, \qquad k = 0, \dots, n-1.$$

Evidently, this is equivalent to the operator equation

$$A_{n,P} u_n := P_n^\sigma A P_n^\sigma u_n = P_n^\sigma f. \tag{2.4}$$

It is clear that (2.4) is in most cases not very suitable for the practical implementation on the computer, since it requires the knowledge of the inner products $\langle A\widetilde{u}_j, \widetilde{u}_k \rangle_\sigma$ and $\langle f, \widetilde{u}_j \rangle_\sigma$ $(j, k = 0, \dots, n-1)$.

The second method, which will play the dominant part in the present paper, is a collocation method, that is, we require (1.1) to be fulfilled only in the collocation points x_{jn}^v, $j = 1, \dots, n$:

$$(Au_n)(x_{jn}^v) = f(x_{jn}^v), \qquad j = 1, \dots, n. \tag{2.5}$$

Of course, this method can only be used if the application of the interpolation operator $\widetilde{L_n^\sigma}$ to the right hand side and to Au_n makes sense (see Section 4). The latter equation can be formulated equivalently by

$$A_{n,L} u_n := \widetilde{L_n^\sigma} A P_n^\sigma u_n = \widetilde{L_n^\sigma} f. \tag{2.6}$$

Remark 2.1 *Of course, (2.5) can also be expressed by $H_n A u_n = H_n f$, where H_n is any interpolation operator with the nodes x_{jn}^v, $j = 1, \dots, n$. We choose $\widetilde{L_n^\sigma}$ (instead of, for instance, L_n^v), since the general theoretical results we are going to apply in the following require that the image space of the approximation operators $A_{n,L}$ coincides with the space $X_n = \operatorname{im} P_n^\sigma$ of the ansatz functions.*

Our main concern is the *applicability* of the approximation methods (2.4) and (2.6) to the original equation. In the following, we are going to define in a somewhat more general situation what we understand by this concept:

Let X be a Banach space. In all what follows, $\mathcal{L}(X, Y)$ means the set of all linear bounded operators between two Banach spaces X and Y, and $\mathcal{L}(X) := \mathcal{L}(X, X)$. Let $\{P_n\} \subset \mathcal{L}(X)$ be a sequence of projections with $P_n \to I$ (strongly), $X_n = \operatorname{im} P_n$, $A \in \mathcal{L}(X)$, and $B_n \in \mathcal{L}(X_n)$, where we require that $B_n P_n$ converges strongly to A.

Definition 2.2 *The projection method with the approximate equations*

$$B_n v_n = P_n f, \qquad v_n \in X_n, \tag{2.7}$$

is said to be applicable to the equation

$$Au = f \tag{2.8}$$

if the following conditions are fulfilled for all $f \in X$:

(i) The equations (2.7) have a unique solution v_n for all sufficiently large n.

(ii) Their solutions v_n converge to a solution of (2.8) for $n \to \infty$.

Definition 2.3 *The sequence $\{B_n\}$ is said to be stable if there is an n_0 such that B_n is invertible for all $n \geq n_0$ and $\sup\limits_{n \geq n_0} \|B_n^{-1} P_n\|_{\mathcal{L}(X)} < \infty$.*

The problem of the applicability of the approximation method (2.7) to (2.8) can be reduced to the stability of $\{B_n\}$ by the following well-known lemma.

Lemma 2.4 ([21], Prop. 1.1) *Let $B_n P_n \to A$ strongly. Then (2.7) is applicable to (2.8) if and only if A is invertible and the sequence $\{B_n\}$ is stable.*

In our concrete situation we have $X = \mathbf{L}^2_\sigma$, $P_n = P_n^\sigma$, $A = aI + bS$ and either $B_n = A_{n,P}|_{X_n}$ in the case of the Galerkin method or $B_n = A_{n,L}|_{X_n}$ in the case of the collocation method. It is clear that $A_{n,P}$ converges strongly to A. The strong convergence of the operators $A_{n,L}$ will be dealt with in Section 4.

Remark 2.5 *As for the collocation method, note that in the following we consider the projection method (2.7) rather than (2.6). If, however, (2.7) is applicable to (1.1) and the right-hand side f satisfies $\|f - \widetilde{L_n^\sigma} f\|_\sigma \to 0 \quad (n \to \infty)$, the solutions u_n of (2.6) converge to the solution u of (1.1).*

Proof. We have

$$\|u_n - u\|_\sigma \leq \left\| B_n^{-1} \widetilde{L_n^\sigma} f - B_n^{-1} B_n P_n^\sigma u \right\|_\sigma + \|P_n^\sigma u - u\|_\sigma$$

$$\leq \|B_n^{-1} P_n^\sigma\|_{\mathcal{L}(\mathbf{L}^2_\sigma)} \left(\left\| \widetilde{L_n^\sigma} f - f \right\|_\sigma + \|Au - B_n P_n^\sigma u\|_\sigma \right) + \|P_n^\sigma u - u\|_\sigma \longrightarrow 0 \quad \blacksquare$$

Remark 2.6 *The preceding remark and its proof show that the convergence of the collocation method is guaranteed in the case of a stable sequence of approximation operators and a suitable right-hand side. On the other hand, for a concrete equation we can also obtain convergence in the case of an unstable sequence $\{B_n\}$ if the approximation of the right-hand side f and of Au is fast enough. Furthermore, in certain cases one can obtain a different (possibly stable) sequence of approximation operators by interpreting (2.5) in a different way (cf. Remark 2.1 and the remarks on Example (C) on page 70).*

3 Banach algebra techniques

3.1 Basic facts

In this subsection we compile some basic facts that will be used later on to investigate the stability problem for our approximation method by Banach algebra techniques.

Definition 3.1 *Let $\mathcal{B}$, $\mathcal{C}$ be unital Banach algebras, $\mathcal{J} \subset \mathcal{B}$ a closed two-sided ideal. A unital homomorphism $\mathcal{W} : \mathcal{B} \to \mathcal{C}$ is called $\mathcal{J}$-lifting if $\mathcal{W}(\mathcal{J})$ is a closed two-sided ideal in $\mathcal{C}$ and $\mathcal{W}|_{\mathcal{J}}$ is an isomorphism between $\mathcal{J}$ and $\mathcal{W}(\mathcal{J})$.*

The following theorem is usually called 'lifting theorem', its first version appears in the paper [49].

Theorem 3.2 (see [21], Theorem 1.8) *Let $\mathcal{B}$, $\mathcal{C}_t$ be unital Banach algebras, where t belongs to an arbitrary index set T, let $\mathcal{J}_t$ be closed two-sided ideals in $\mathcal{B}$ and let $\mathcal{W}_t : \mathcal{B} \to \mathcal{C}_t$ be $\mathcal{J}_t$-lifting homomorphisms. By $\mathcal{J}$ we denote the smallest closed two-sided ideal in $\mathcal{B}$ that contains all $\mathcal{J}_t$, $t \in T$. Then an element $b \in \mathcal{B}$ is invertible in $\mathcal{B}$ if and only if $\mathcal{W}_t(b)$ is invertible in $\mathcal{C}_t$ for all $t \in T$ and the coset $b + \mathcal{J}$ is invertible in the quotient algebra $\mathcal{B}/\mathcal{J}$.*

The invertibility of the coset $b + \mathcal{J}$ is usually investigated with the help of local principles.

Definition 3.3 *Let $\mathcal{B}$ be a unital Banach algebra. A subset $M \subset \mathcal{B}$ is called a localizing class if $0 \notin M$ and if for all a_1, $a_2 \in M$ there exists an element $a \in M$ such that*

$$aa_j = a_j a = a \qquad (j = 1, 2).$$

In the following let M be a localizing class. Two elements x, $y \in \mathcal{B}$ are called M-equivalent (in symbols: $x \overset{M}{\sim} y$), if

$$\inf_{a \in M} \|a(x - y)\| = \inf_{a \in M} \|(x - y)a\| = 0.$$

Further, $x \in \mathcal{B}$ is called M-invertible if there exist a_1, $a_2 \in M$, z_1, $z_2 \in \mathcal{B}$ such that

$$z_1 x a_1 = a_1, \qquad a_2 x z_2 = a_2.$$

(Note that an invertible element is also M_τ-invertible.) A system $\{M_\tau\}_{\tau \in T}$ of localizing classes (T is an arbitrary index set) is said to be covering, if for each system $\{a_\tau\}_{\tau \in T}$, $a_\tau \in M_\tau$, there exists a finite subsystem $a_{\tau_1}, \dots, a_{\tau_n}$ such that $a_{\tau_1} + \dots + a_{\tau_n}$ is invertible in $\mathcal{B}$.

Now we can formulate the local principle of Gohberg and Krupnik:

Theorem 3.4 ([20], Theorem XII.1.1) *Let $\mathcal{B}$ be a unital Banach algebra, and let $\{M_\tau\}_{\tau \in T}$ be a covering system of localizing classes in $\mathcal{B}$, $x \in \mathcal{B}$ and $x \overset{M_\tau}{\sim} x_\tau$ for all $\tau \in T$. Further, assume that x commutes with all elements from $\bigcup_{\tau \in T} M_\tau$. Then x is invertible in $\mathcal{B}$ if and only if x_τ is M_τ-invertible for all $\tau \in T$.*

Another local principle is due to Allan and Douglas:

Theorem 3.5 ([6], Theorem 1.34) *Let $\mathcal{B}$ be a unital Banach algebra and $\mathcal{C} \subset \mathcal{B}$ a closed central subalgebra (that is, all elements of $\mathcal{C}$ commute with all elements of $\mathcal{B}$) that contains the unit element. For every maximal ideal τ of $\mathcal{C}$ we introduce the local ideal J_τ as the smallest closed two-sided ideal of $\mathcal{B}$ that contains τ. Then*

(i) An element $x \in \mathcal{B}$ is invertible in $\mathcal{B}$ if and only if the cosets $x + J_\tau$ are invertible in $\mathcal{B}/J_\tau$ for all τ.

(ii) $\bigcap_\tau J_\tau$ is contained in the radical of $\mathcal{B}$.

Remark 3.6 *We recall that the radical of an algebra is defined as the intersection of all maximal left-sided (or right-sided) ideals. If $\mathcal{B}$ is a C^*-algebra, the radical of $\mathcal{B}$ and therefore also the intersection of the local ideals is trivial.*

Remark 3.7 (see [44], proof of Theorem 1.21) *If $\mathcal{B}$, $\mathcal{C}$ are C^*-algebras, there is a close relation between the two local principles. Let $M(\mathcal{C})$ denote the maximal ideal space of $\mathcal{C}$. For $\tau \in M(\mathcal{C})$ we define*

$$M_\tau := \{a \in \mathcal{C} : 0 \le (Ga)(s) \le 1,\ (Ga)(s) \equiv 1 \text{ in some neighbourhood of } \tau\},$$

where $G : \mathcal{C} \to \mathbf{C}(M(\mathcal{C}))$ denotes the Gelfand map. Then $\{M_\tau\}_{\tau\in M(\mathcal{C})}$ forms a covering system of localizing classes in $\mathcal{B}$, and the local ideals occurring in the principle of Allan and Douglas can be described by $J_\tau = \{x \in \mathcal{B} : x \overset{M_\tau}{\sim} 0\}$.

3.2 Application to stability analysis

We want to investigate the applicability of the approximation method (2.7) to equation (2.8). In all what follows we assume that $B_n P_n$ converges strongly to A, which allows us to reduce the problem to the question of whether $\{B_n\}$ is stable. Furthermore, we specify X to be a Hilbert space.

By $\mathcal{E}$ we denote the set of all operator sequences $\{B_n P_n\}$, where $B_n \in \mathcal{L}(X_n)$ and $\sup_n \|B_n P_n\|_{\mathcal{L}(X)} < \infty$. Endowed with componentwise algebraic operations and the norm $\|\{A_n\}\|_{\mathcal{E}} = \sup_n \|A_n\|_{\mathcal{L}(X)}$, $\mathcal{E}$ becomes a C^*-algebra. The set $\mathcal{N} := \{\{C_n\} \in \mathcal{E} : \|C_n\| \to 0\}$ is a closed ideal in $\mathcal{E}$. In the sequel we will use the notation $\mathcal{GB}$ for the set of all invertible elements of a Banach algebra $\mathcal{B}$. The following well-known result identifies the question of stability with an invertibility problem.

Lemma 3.8 ([21], Proposition 1.2) *A sequence $\{A_n\} \in \mathcal{E}$ is stable if and only if $\{A_n\} + \mathcal{N} \in \mathcal{G}(\mathcal{E}/\mathcal{N})$.*

We introduce a further family of operators $W_n \in \mathcal{L}(X)$, where we assume that $W_n = W_n^*$ converges weakly to 0, $W_n P_n = W_n$ and $W_n^2 = P_n$. Let $\mathcal{A}$ denote the set of all sequences $\{A_n\} \in \mathcal{E}$ for which A_n, A_n^*, $\widetilde{A_n} := W_n A_n W_n$ and $\widetilde{A}_n^*$ are strongly convergent. (As for the transition to the adjoint operator, note that we consider A_n as an operator acting on the whole space X here.)

Lemma 3.9 *The set $\mathcal{A}$ is a C^*-algebra.*

Proof. Evidently, $\mathcal{A}$ is an algebra. If $\{A_n\} \in \mathcal{A}$, then so is $\{A_n^*\}$ (note that $\{W_n A_n^* W_n\} = \{(W_n A_n W_n)^*\}$). Let $\{A^{(n)}\}$ be a fundamental sequence in $\mathcal{A}$, where $A^{(n)} = \{A_k^{(n)}\}_{k=1}^\infty$. If $\varepsilon > 0$ is given, we have

$$\|A_k^{(n)} - A_k^{(m)}\| < \varepsilon$$

for all k if m, n are large enough. Hence, there is a sequence $\{A_k\} \in \mathcal{L}(X)$ such that $\|A_k^{(n)} - A_k\| \to 0 \quad (n \to \infty)$ uniformly with respect to k. If we choose $x \in X$, we can estimate

$$\|(A_k - A_l)x\| \leq \|A_k - A_k^{(n)}\| \, \|x\| + \|A_l - A_l^{(n)}\| \, \|x\| + \|(A_k^{(n)} - A_l^{(n)})x\|.$$

The first two terms can be made arbitrarily small by the choice of n, and the third one goes to zero if n is fixed and $k, l \to \infty$, since $\{A_k^{(n)}\}_{k=1}^\infty \in \mathcal{A}$. Hence, A_k is strongly convergent. The sequences A_k^*, $\widetilde{A}_k$ and $\widetilde{A}_k^*$ are treated in the same way (note that $\|\widetilde{A}_k^{(n)} - \widetilde{A}_k^{(m)}\| \leq \text{const } \|A_k^{(n)} - A_k^{(m)}\|$ because of the weak convergence of W_n). ■

In the following, the coset $\{A_n\} + \mathcal{N}$ of a sequence $\{A_n\} \in \mathcal{E}$ will be denoted by $\widehat{\{A_n\}}$. The ideal $\mathcal{N}$ is contained in $\mathcal{A}$, and thus $\widehat{\mathcal{A}} := \mathcal{A}/\mathcal{N}$ is a C^*-subalgebra of $\widehat{\mathcal{E}} := \mathcal{E}/\mathcal{N}$ and is therefore inverse-closed (that means $\widehat{\mathcal{A}} \cap \mathcal{G}\widehat{\mathcal{E}} = \mathcal{G}\widehat{\mathcal{A}}$, see [44, 1.13]). Hence, the stability of $\{A_n\} \in \mathcal{A}$ is equivalent to $\widehat{\{A_n\}} \in \mathcal{G}\widehat{\mathcal{A}}$.

In all what follows, $\mathcal{K}(X, Y)$ denotes the space of all compact linear operators between two Banach spaces X and Y. If $X = Y$, we briefly write $\mathcal{K}(X)$ instead of $\mathcal{K}(X, X)$.

Lemma 3.10 ([44], 1.1.h) *Let X, Y, Z, V be Banach spaces, $\{A_n\} \subset \mathcal{L}(Z, V)$, $\{B_n\} \subset \mathcal{L}(X, Y)$ such that $A_n \to A \in \mathcal{L}(Z, V)$, $B_n^* \to B^* \in \mathcal{L}(Y^*, X^*)$ strongly. If $K \in \mathcal{K}(Y, Z)$, then $\|A_n K B_n - AKB\|_{\mathcal{L}(X,V)} \to 0 \quad (n \to \infty)$. If $K \in \mathcal{K}(Y, Z)$ and $B_n \rightharpoonup B \in \mathcal{L}(X, Y)$ (weakly), then $KB_n \to KB$ strongly.*

We define

$$\mathcal{J}_0 := \left\{ \widehat{\{P_n K P_n\}} : K \in \mathcal{K}(X) \right\}, \quad \mathcal{J}_1 := \left\{ \widehat{\{W_n K W_n\}} : K \in \mathcal{K}(X) \right\}.$$

By Lemma 3.10, $\mathcal{J}_0$, $\mathcal{J}_1$ are subsets of $\widehat{\mathcal{A}}$.

Lemma 3.11 ([49], Satz 2) *$\mathcal{J}_0$, $\mathcal{J}_1$ are closed ideals in $\widehat{\mathcal{A}}$, and the smallest closed ideal containing $\mathcal{J}_0$ and $\mathcal{J}_1$ equals*

$$\mathcal{J} = \{\{P_n K_1 P_n + W_n K_2 W_n\} + \mathcal{N} : K_1, \, K_2 \in \mathcal{K}(X)\}.$$

Proof. We show that $\mathcal{I}_0 := \{\{P_n K P_n + C_n\} : K \in \mathcal{K}(X), \, \|C_n\| \to 0\}$ is a closed ideal in $\mathcal{A}$, whence the corresponding property of $\mathcal{J}_0$ in $\widehat{\mathcal{A}}$ follows immediately. Obviously, $\mathcal{I}_0$ is a linear space. Let $\{B^n\} \subset \mathcal{I}_0$ be a fundamental sequence, $B^n = \{A_k^{(n)}\}_{k=1}^\infty$, $A_k^{(n)} = P_k T^{(n)} P_k + C_k^{(n)}$, where $T^{(n)} \in \mathcal{K}(X)$, $\|C_k^{(n)}\| \to 0 \quad (k \to \infty)$. We have $A_k^{(n)} \to T^{(n)}$ (strongly), and hence for $\varepsilon > 0$ the relation $\|T^{(n)} - T^{(m)}\| \leq \sup_k \|A_k^{(n)} - A_k^{(m)}\| < \varepsilon$

holds for n, m large enough. Therefore, $T^{(n)}$ converges uniformly to some $T \in \mathcal{K}(X)$. Besides, we can estimate

$$\|C_k^{(n)} - C_k^{(m)}\| \leq \|P_k(T^{(n)} - T^{(m)})P_k\| + \|A_k^{(n)} - A_k^{(m)}\| < \varepsilon$$

for sufficiently large m, n independently of k. Thus, there exists a sequence $\{C_k\}$ with $\|C_k^{(n)} - C_k\| \to 0 \quad (n \to \infty)$, and since $\|C_k\| \leq \|C_k^{(n)} - C_k\| + \|C_k^{(n)}\|$, we have $\{C_k\} \in \mathcal{N}$. If we put $B := \{P_kTP_k + C_k\}$, we have

$$\|B^n - B\|_{\mathcal{A}} \leq \text{const } \|T^{(n)} - T\| + \sup_k \|C_k^{(n)} - C_k\| \longrightarrow 0 \qquad (n \to \infty),$$

which proves the closedness of $\mathcal{I}_0$. To show (for instance) that $\mathcal{I}_0$ is a left ideal, let $\{A_k\} \in \mathcal{A}$, $A_k \to A$, $\{B_k\} = \{P_kTP_k + C_k\} \in \mathcal{I}_0$. Then

$$\{A_k\}\{B_k\} = \{P_kA_kTP_k + A_kC_k\} = \{P_kATP_k + \underbrace{P_k(A_k - A)TP_k + A_kC_k}_{\in \mathcal{N}}\} \in \mathcal{I}_0$$

(cf. Lemma 3.10). The proof for $\mathcal{J}_1$, $\mathcal{J}$ is analogous. Obviously, $\mathcal{J}$ is contained in every ideal that contains $\mathcal{J}_0$ and $\mathcal{J}_1$, which completes the proof. ■

For $\widehat{\{A_n\}} \in \widehat{\mathcal{A}}$ we define

$$\mathcal{W}_0\widehat{\{A_n\}} := \text{s-}\lim_{n\to\infty} A_n, \qquad \mathcal{W}_1\widehat{\{A_n\}} := \text{s-}\lim_{n\to\infty} W_nA_nW_n,$$

where $\text{s-}\lim A_n$ denotes the strong limit of the operator sequence A_n. Note that $\mathcal{W}_0$, $\mathcal{W}_1$ are correctly defined. Evidently, $\mathcal{W}_i \quad (i = 0, 1)$ are continuous *-homomorphisms from $\widehat{\mathcal{A}}$ into $\mathcal{L}(X)$ (remember that $\|\text{s-}\lim A_n\| \leq \liminf \|A_n\|$). Further, it is easy to see that $\mathcal{W}_i|_{\mathcal{J}_i}$ is an isomorphism between $\mathcal{J}_i$ and the ideal $\mathcal{K}(X)$ of all compact linear operators on X, in other words, $\mathcal{W}_i$ is $\mathcal{J}_i$-lifting. If we apply Theorem 3.2 with $T = \{0, 1\}$ to this situation, we obtain the original version of the lifting theorem:

Theorem 3.12 ([49], Satz 3) *Let $\{A_n\} \in \mathcal{A}$, $A_n \to A$, $\widetilde{A_n} \to \widetilde{A}$ strongly. Then $\{A_n\}$ is stable if and only if $A, \widetilde{A} \in \mathcal{GL}(X)$ and $\widehat{\{A_n\}} + \mathcal{J} \in \mathcal{G}(\widehat{\mathcal{A}}/\mathcal{J})$.*

Remark 3.13 *Instead of the third condition of the preceding theorem we will verify the equivalent formulation $\{A_n\} + \mathcal{I} \in \mathcal{G}(\mathcal{A}/\mathcal{I})$, with the ideal $\mathcal{I} = \{\{P_nK_1P_n + W_nK_2W_n + C_n\} : K_1, K_2 \in \mathcal{K}(X), \|C_n\| \to 0\} \subset \mathcal{A}$.*

In the sequel we will concentrate on considering the stability of the sequence $\{A_n\} = \{A_{n,L}\}$ related to the collocation method. The role of the W_n will be played by the operators W_n^σ defined by

$$W_n^\sigma u := \sum_{k=0}^{n-1} \langle u, \widetilde{u}_{n-1-k}\rangle_\sigma \widetilde{u}_k,$$

which obviously have the required properties. In Section 4 we will show that $\{A_n\} \in \mathcal{A}$ and compute the strong limit $\widetilde{A} = \mathcal{W}_1\{A_n\}$, and Section 5 is mainly dedicated to investigating the invertibility of the coset $\{A_n\} + \mathcal{I}$.

As for the Galerkin method, we will not directly apply the concepts of the present section (some partial results in this direction can be found in [53]), since one can transform the problem to a Galerkin method for operators on the unit circle, for which stability conditions are already known.

4 Strong convergence of the operator sequences

The results of this section are slight generalizations of those obtained in [32]. The strong convergence of $A_n = \widetilde{L_n^\sigma}(aI + bS)P_n^\sigma$ and of A_n^* will be shown for Riemann integrable coefficients. For $\widetilde{A_n}$ and $\widetilde{A_n}^*$ we will require the piecewise continuity of b.

4.1 Strong convergence of A_n

First we give sufficient conditions for the weighted interpolation polynomial $\widetilde{L_n^\sigma} f$ to converge in the $\mathbf{L}^2_\sigma$-norm. For this end, we provide some material from [16] and [43] concerning the convergence of Gaussian quadrature rules and Lagrangian interpolation operators.

Consider the Jacobi weight v. Let $x_{kn}^v \quad (k = 1, \dots, n)$ be the zeros of the orthogonal polynomial of degree n related to v, and L_n^v the Lagrangian interpolation operator with respect to x_{kn}^v. By $\mathbf{Q}_n^v$ we denote the Gaussian quadrature rule

$$\mathbf{Q}_n^v f := \int_{-1}^1 (L_n^v f)(x) v(x) \, \mathrm{d}x = \sum_{k=1}^n A_{kn}^v f(x_{kn}^v).$$

Definition 4.1 *Let $v = v^{\gamma,\delta}$. For $\varepsilon > 0$, denote by $\mathbf{R}_v^\varepsilon$ the set of all complex-valued functions $f : (-1,1) \to \mathbb{C}$ that are locally Riemann integrable and for which $v^{\gamma+1-\varepsilon,\delta+1-\varepsilon} f$ is bounded. $\mathbf{R}_v^\varepsilon$ becomes a Banach space if we introduce the norm*

$$\|f\|_{\mathbf{R}_v^\varepsilon} = \left\| v^{\gamma+1-\varepsilon,\delta+1-\varepsilon} f \right\|_\infty .$$

Lemma 4.2 (cf. [16], Satz III.1.6b, see also [32], Lemma 4.3) *Assume that f is bounded on every compact subinterval of $(-1,1)$ and the (improper) Riemann integral $\int_{-1}^1 f(x)v(x)\,\mathrm{d}x$ exists. Suppose that there exist functions g_{-1}, g_1 for which $g_j^{(2\nu)}(x) \geq 0$ for all $x \in (-1,1)$ and $\nu = 0,1,2,\dots$ and $\int_{-1}^1 g_j(x)v(x)\,\mathrm{d}x < \infty$, $j = \pm 1$. Furthermore, let*

$$\lim_{x\to -1+0} \frac{f(x)}{g_{-1}(x)} = \lim_{x \to 1-0} \frac{f(x)}{g_1(x)} = 0.$$

Then $\mathbf{Q}_n^v f \to \int_{-1}^1 f(x)v(x)\,\mathrm{d}x \quad (n \to \infty)$.

Corollary 4.3 (cf. [16], Satz III.1.6b) *If $v = v^{\gamma,\delta}$ with γ, δ satisfying (1.4) we can choose $g_{-1} = (1+x)^{-1-\delta+\varepsilon}$, $g_1(x) = (1-x)^{-1-\gamma+\varepsilon}$ with some $\varepsilon > 0$. Hence, the Gaussian quadrature rule converges if $f \in \mathbf{R}_v^\varepsilon$, $\varepsilon > 0$.*

Lemma 4.4 (cf. [16], Satz III.2.1) *Assume that $|f|^2 \in \mathbf{R}_v^\varepsilon$ for some $\varepsilon > 0$. Then the relation $\lim_{n\to\infty} \|L_n^v f - f\|_v = 0$ holds.*

Proof. According to [16], Satz III.4.3, the polynomials are dense in $\mathbf{L}^2_v$. (This remains true in the complex case, since real and imaginary part can be approximated separately.) Let $\eta > 0$, and let p be a polynomial with $\|p - f\|_v < \eta$. For $n > \deg p$ we have

$$\begin{aligned}\|f - L^v_n f\|^2_v &\leq 2(\|f - p\|^2_v + \|L^v_n(p - f)\|^2_v)\\ &< 2(\eta^2 + \mathbf{Q}^v_n(|p - f|^2))\end{aligned}$$

(note that the Gaussian quadrature rule with n nodes is exact for polynomials of degree less than $2n$). Since $\varepsilon_1 < \varepsilon_2$ implies $\mathbf{R}^{\varepsilon_2}_v \subset \mathbf{R}^{\varepsilon_1}_v$, we can assume without loss of generality that $\varepsilon \leq \min\{1 + \gamma, 1 + \delta\}$. Then we have the relation

$$|p(x) - f(x)|^2 \leq 2(|p(x)|^2 + |f(x)|^2) \leq \text{const}\, v^{-1-\gamma+\varepsilon,-1-\delta+\varepsilon},$$

that is $|p - f|^2 \in \mathbf{R}^{\varepsilon}_v$. Corollary 4.3 now yields

$$\mathbf{Q}^v_n(|p - f|^2) \to \int_{-1}^{1} |p(x) - f(x)|^2 v(x)\,\mathrm{d}x < \eta^2. \quad \blacksquare$$

Corollary 4.5 *Let $\sigma = v^{\alpha,\beta}$. If f is locally Riemann integrable on $(-1, 1)$ and*

$$|f(x)| \leq \text{const}\,(1 - x)^{(-1-\alpha)/2+\varepsilon}(1 + x)^{(-1-\beta)/2+\varepsilon}$$

with some $\varepsilon > 0$, then

$$\|\widetilde{L^\sigma_n} f - f\|_\sigma \to 0 \qquad (n \to \infty).$$

Proof. Using the isometric isomorphism $w_{v,\sigma^{-1}}I : \mathbf{L}^2_v \to \mathbf{L}^2_\sigma$, we have

$$\|\widetilde{L^\sigma_n} f - f\|_\sigma = \|L^v_n w^{-1}_{v,\sigma^{-1}} f - w^{-1}_{v,\sigma^{-1}} f\|_v.$$

Now the assertion follows from Lemma 4.4. $\blacksquare$

In the sequel we will investigate the behaviour of the operators A_n. First we do this for the multiplication operator $A = aI$ separately.

Proposition 4.6 *Let a be Riemann integrable. Then $\widetilde{L^\sigma_n} a P^\sigma_n \to aI$ strongly on $\mathbf{L}^2_\sigma$. Furthermore, we have the estimation $\|\widetilde{L^\sigma_n} a P^\sigma_n\|_{\mathcal{L}(\mathbf{L}^2_\sigma)} \leq \|a\|_\infty$.*

Proof. First we show the convergence on the dense subset span $\{\widetilde{u}_m\}^\infty_{m=0}$. Clearly, the functions $a\widetilde{u}_m$ satisfy the conditions of Corollary 4.5. Thus, for $n > m$ we have $\|A_n\widetilde{u}_m - A\widetilde{u}_m\|_\sigma = \|\widetilde{L^\sigma_n} a\widetilde{u}_m - a\widetilde{u}_m\|_\sigma \to 0$. For showing the uniform boundedness, let $u \in \mathbf{L}^2_\sigma$ and write $P^\sigma_n u = w_{v,\sigma^{-1}} q_n$ with a polynomial q_n of degree less than n. If we again note the exactness of the Gaussian quadrature rule, we have

$$\begin{aligned}\|\widetilde{L^\sigma_n} a P^\sigma_n u\|^2_\sigma &= \|L^v_n a q_n\|^2_v = \mathbf{Q}^v_n(|a q_n|^2)\\ &\leq \|a\|^2_\infty \mathbf{Q}^v_n(|q_n|^2) = \|a\|^2_\infty \|q_n\|^2_v = \|a\|^2_\infty \|P^\sigma_n u\|^2_\sigma \leq \|a\|^2_\infty \|u\|^2_\sigma.\end{aligned}$$

Now the assertion follows from the Banach-Steinhaus theorem. $\blacksquare$

Let ϱ be a Jacobi weight and $\mu \in (0, 1)$. By $\mathbf{H}^\mu_0(\varrho)$ we denote the Banach space of all functions f for which $\varrho f \in \mathbf{C}^{0,\mu}[-1, 1]$ and $(\varrho f)(\pm 1) = 0$. The norm in this space is defined by $\|f\|_{\mathbf{H}^\mu_0(\varrho)} := \|\varrho f\|_{\mathbf{C}^{0,\mu}}$.

Lemma 4.7 ([20], Theorem I.6.2) *Let $\varrho = v^{s,t}$, $\mu \in (0,1)$ and $\mu < s,t < \mu+1$. Then the Cauchy singular integral operator S defined by (1.3) is bounded on $\mathbf{H}_0^\mu(\varrho)$.*

In what follows, by "const" we will denote positive constants that may have different values at different places.

Proposition 4.8 *The operators $\widetilde{L_n^\sigma} S P_n^\sigma$ converge strongly to S on* span $\{\widetilde{u}_m\}_{m=0}^\infty$.

Proof. Let $\varrho = v^{s,t}$, where we choose s,t such that

$$\max\left\{0, \frac{\alpha}{2} - \frac{\gamma}{2}\right\} < s < \frac{1+\alpha}{2}, \qquad \max\left\{0, \frac{\beta}{2} - \frac{\delta}{2}\right\} < t < \frac{1+\beta}{2},$$

and let

$$0 < \mu < \min\left\{s, t, s + \frac{\gamma}{2} - \frac{\alpha}{2}, t + \frac{\delta}{2} - \frac{\beta}{2}\right\}.$$

Then we have $\mu < s,t < \mu+1$ and $\widetilde{u}_m \in \mathbf{H}_0^\mu(\varrho)$. The latter relation follows from

$$(\varrho\widetilde{u}_m)(x) = p_m^v(x)(1-x)^{s+(\gamma-\alpha)/2}(1+x)^{t+(\delta-\beta)/2},$$

the exponents being greater than μ. By Lemma 4.7, we also have $S\widetilde{u}_m \in \mathbf{H}_0^\mu(\varrho)$, which means that $h := \varrho S\widetilde{u}_m \in \mathbf{C}^{0,\mu}$. Thus, we can estimate

$$|(S\widetilde{u}_m)(x)| = |(\varrho^{-1}h)(x)| = |h(x)(1-x)^{-s}(1+x)^{-t}|$$

$$\leq \text{const } (1-x)^{-(1+\alpha)/2+\varepsilon}(1+x)^{-(1+\beta)/2+\varepsilon}$$

if $\varepsilon > 0$ is small enough. Corollary 4.5 now gives the assertion. ■

In all what follows we exclude the cases $\alpha = \gamma$ and $\beta = \delta$ since they bring about some technical difficulties in the proofs that we could only partially overcome. Some remarks concerning these cases will be given in Subsection 4.3.

If, as we assumed, $\alpha \neq \gamma$ and $\beta \neq \delta$, we can choose $\widetilde{b} \equiv 1$ and $\widetilde{a} \in \mathbf{C}^1$ in Proposition 1.1 (cf. the notations used there) such that $\lambda := \frac{\gamma-\alpha}{2} - g(1)$ and $\nu := \frac{\delta-\beta}{2} + g(-1)$ are integers, whence we get the mapping properties described in Proposition 1.1 for the operator $(\widetilde{a}I + iS)w_{v,\sigma^{-1}}wI$ with some positive function w, $w \in \mathbf{C}^{0,\eta}$ for all $\eta \in (0,1)$. Furthermore, we can always achieve $\kappa \geq -1$ (we need the latter relation to guarantee the exactness of the Gaussian quadrature rule): Evidently, we always have $g(x) \in (-1,0)$.

- In case $\alpha_0 := \frac{\gamma-\alpha}{2} < 0$, $\beta_0 := \frac{\delta-\beta}{2} < 0$ we choose $g(1) = \alpha_0, g(-1) = -1-\beta_0$ and have $\kappa = 1$.

- If $\alpha_0, \beta_0 > 0$, let $g(1) = \alpha_0 - 1$, $g(-1) = -\beta_0$, whence $\kappa = -1$.

- Finally, let $g(1) = \alpha_0$, $g(-1) = -\beta_0$ if $\alpha_0 < 0 < \beta_0$, and $g(1) = \alpha_0 - 1$, $g(-1) = -1 - \beta_0$ if $\beta_0 < 0 < \alpha_0$, both of which results in $\kappa = 0$.

Lemma 4.9 ([43], Theorem 9.25) *Let v, v^* be Jacobi weights with $vv^* \in \mathbf{L}^1$, and let l be a fixed positive integer. If q is a polynomial with $\deg q \le ln$, then we have*

$$\sum_{k=1}^{n} A_{kn}^v |q(x_{kn}^v)| v^*(x_{kn}^v) \le \text{const} \int_{-1}^{1} |q(x)| v(x) v^*(x)\, dx,$$

the constant being independent of n and q.

Our proof of the uniform boundedness of $\widetilde{L_n^\sigma} S P_n^\sigma$ will be based on the following decomposition of the operator S:

$$S = i\tilde{a}I - iw^{-1}(\tilde{a}I + iS)wI + w^{-1}(wS - SwI). \tag{4.1}$$

Now we are going to estimate the three summands of $\widetilde{L_n^\sigma} S P_n^\sigma$ separately.

By virtue of Proposition 4.6 we have $\|\widetilde{L_n^\sigma} \tilde{a} P_n^\sigma\| \le \|\tilde{a}\|_\infty$, so we are done with the first term.

Proposition 4.10 *We have the estimation*

$$\|\widetilde{L_n^\sigma} w^{-1}(\tilde{a}I + iS) w P_n^\sigma\|_{\mathcal{L}(\mathbf{L}_\sigma^2)} \le \text{const} \; \|w^{-1}\|_\infty \, \|\tilde{a}wI + iSwI\|_{\mathcal{L}(\mathbf{L}_\sigma^2)}.$$

Proof. According to Proposition 1.1, $q_{n-\kappa} := (\tilde{a}I + iS)wP_n^\sigma u$ is a polynomial of degree less than $n-\kappa$ for all $u \in \mathbf{L}_\sigma^2$. Now we have

$$\left\|\widetilde{L_n^\sigma} w^{-1}(\tilde{a}I + iS) w P_n^\sigma u\right\|_\sigma^2 = \|L_n^v w^{-1}(w_{v,\sigma^{-1}})^{-1} q_{n-\kappa}\|_v^2$$

$$= \sum_{k=1}^{n} A_{kn}^v |w^{-1}(x_{kn}^v)|^2 (w_{v,\sigma^{-1}}(x_{kn}^v))^{-2} |q_{n-\kappa}(x_{kn}^v)|^2$$

$$\le \|w^{-1}\|_\infty^2 \sum_{k=1}^{n} A_{kn}^v (w_{v,\sigma^{-1}}(x_{kn}^v))^{-2} |q_{n-\kappa}(x_{kn}^v)|^2$$

$$\le \text{const} \; \|w^{-1}\|_\infty^2 \int_{-1}^{1} |q_{n-\kappa}(x)|^2 (w_{v,\sigma^{-1}}(x))^{-2} v(x)\, dx$$

$$= \text{const} \; \|w^{-1}\|_\infty^2 \; \|q_{n-\kappa}\|_\sigma^2$$

$$\le \text{const} \; \|w^{-1}\|_\infty^2 \; \|\tilde{a}wI + iSwI\|_{\mathcal{L}(\mathbf{L}_\sigma^2)}^2 \; \|u\|_\sigma^2 ,$$

where we used Lemma 4.9 with $v^* = w_{v,\sigma^{-1}}^{-2}$. ∎

Lemma 4.11 ([8], Lemma 2.4) *Let $\alpha, \beta, s > -1$. Then we have*

$$\int_{-1}^{1} |t - x|^s (1-t)^{-\alpha} (1+t)^{-\beta}\, dt \le \text{const} \; (1-x)^{-\alpha^+} (1+x)^{-\beta^+}$$

for $-1 < x < 1$, where $\alpha^+ := \max\{\alpha, 0\}$, $\beta^+ := \max\{\beta, 0\}$.

The following lemma is a slight generalization of [25, Lemma 2.3].

Lemma 4.12 *Let* $\sigma = v^{\alpha,\beta}$ *and* $0 < \zeta < m := \frac{1-\max\{\alpha,\beta,0\}}{2}$. *Then*

$$\int_{-1}^{1} \left| \frac{1}{|t-x|^{\zeta}} - \frac{1}{|t-y|^{\zeta}} \right|^2 \sigma^{-1}(t)\, dt \leq \text{const } |x-y|^{2\lambda}$$

for all $\lambda \in (0,1)$ *with* $\lambda + \zeta < m$.

Proof. Let $\lambda_0 := \lambda + \zeta$. We have

$$\begin{aligned} |t-x|^{-\zeta} - |t-y|^{-\zeta} &= \frac{|t-x|^{\lambda_0-\zeta} - |t-y|^{\lambda_0-\zeta}}{|t-x|^{\lambda_0}} + \frac{|t-x|^{\lambda_0-\zeta} - |t-y|^{\lambda_0-\zeta}}{|t-y|^{\lambda_0}} \\ &+ \frac{|t-y|^{2\lambda_0-\zeta} - |t-x|^{2\lambda_0-\zeta}}{|t-x|^{\lambda_0}|t-y|^{\lambda_0}}. \end{aligned}$$

Hence, we can estimate

$$\begin{aligned} &\left||t-x|^{-\zeta} - |t-y|^{-\zeta}\right|^2 \\ &\leq 3\left(|t-x|^{\lambda_0-\zeta} - |t-y|^{\lambda_0-\zeta}\right)^2 \left(\frac{1}{|t-x|^{2\lambda_0}} + \frac{1}{|t-y|^{2\lambda_0}}\right) \\ &\quad + 3\frac{\left(|t-y|^{2\lambda_0-\zeta} - |t-x|^{2\lambda_0-\zeta}\right)^2}{|t-x|^{2\lambda_0}|t-y|^{2\lambda_0}} \\ &\leq \text{const } \left[|x-y|^{2(\lambda_0-\zeta)}\left(\frac{1}{|t-x|^{2\lambda_0}} + \frac{1}{|t-y|^{2\lambda_0}}\right) + \frac{|x-y|^{4\lambda_0-2\zeta}}{|t-x|^{2\lambda_0}|t-y|^{2\lambda_0}}\right] \\ &= \text{const } |x-y|^{2\lambda}\left[\frac{1}{|t-x|^{2\lambda_0}} + \frac{1}{|t-y|^{2\lambda_0}} + \left|\frac{1}{t-x} - \frac{1}{t-y}\right|^{2\lambda_0}\right] \\ &\leq \text{const } |x-y|^{2\lambda}\left(\frac{1}{|t-x|^{2\lambda_0}} + \frac{1}{|t-y|^{2\lambda_0}}\right). \end{aligned}$$

Obviously, we have $\sigma^{-1} \in \mathbf{L}^p$ for $p < \frac{1}{\max\{\alpha,\beta,0\}}$. Then for the adjoint exponent $q = \frac{p}{p-1}$ the relation $q > \frac{1}{1-\max\{\alpha,\beta,0\}} = \frac{1}{2m}$ holds. Since $\lambda_0 < m$, we can choose p such that $2\lambda_0 q < 1$. Hence, the Hölder inequality gives

$$\int_{-1}^{1} \frac{\sigma^{-1}(t)}{|t-x|^{2\lambda_0}}\, dt \leq \|\sigma^{-1}\|_{\mathbf{L}^p} \left(\int_{-1}^{1} \frac{dt}{|t-x|^{2\lambda_0 q}}\right)^{\frac{1}{q}} \leq \text{const}$$

independently of x (compare Lemma 4.11), and the lemma is proved if we apply this relation to the preceding estimation. ■

Lemma 4.13 *Let* $\sigma = v^{\alpha,\beta}$ *and* $w \in \mathbf{C}^{0,\eta}$ *with* $\eta > \frac{1+\max\{\alpha,\beta,0\}}{2}$. *Then the operator* $K := wS - SwI$ *belongs to* $\mathcal{K}(\mathbf{L}^2_\sigma, \mathbf{C}^{0,\lambda})$ *for some* $\lambda > 0$.

Proof. Choose $1-\eta < \zeta < \frac{1-\max\{\alpha,\beta,0\}}{2}$ and put $k(t,x) := \frac{w(t)-w(x)}{t-x}|t-x|^\zeta$. If $\lambda \le \eta-(1-\zeta)$, we have $k \in \mathbf{C}^{0,\lambda}$ in both variables, uniformly with respect to the other ([42], §5). Moreover, we require that λ satisfies $\zeta+\lambda < \frac{1-\max\{\alpha,\beta,0\}}{2}$. Let now $u \in \mathbf{L}^2_\sigma$. We can write

$$\begin{aligned} |(Ku)(x)-(Ku)(y)| &\le \frac{1}{\pi}\int_{-1}^{1}\left|\frac{k(t,x)}{|t-x|^\zeta}-\frac{k(t,y)}{|t-y|^\zeta}\right| |u(t)|\,dt \\ &\le \frac{1}{\pi}\left(\int_{-1}^{1}\left|\frac{k(t,x)}{|t-x|^\zeta}-\frac{k(t,y)}{|t-y|^\zeta}\right|^2 \sigma^{-1}(t)\,dt\right)^{\frac{1}{2}} \|u\|_\sigma , \end{aligned}$$

and, using Lemma 4.12, the following estimation holds:

$$\begin{aligned} &\int_{-1}^{1}\left|\frac{k(t,x)}{|t-x|^\zeta}-\frac{k(t,y)}{|t-y|^\zeta}\right|^2 \sigma^{-1}(t)\,dt \\ &\le \int_{-1}^{1}\left(\left|\frac{k(t,x)-k(t,y)}{|t-x|^\zeta}\right| + |k(t,y)|\left|\frac{1}{|t-x|^\zeta}-\frac{1}{|t-y|^\zeta}\right|\right)^2 \sigma^{-1}(t)\,dt \\ &\le 2\int_{-1}^{1}\left(\left|\frac{k(t,x)-k(t,y)}{|t-x|^\zeta}\right|^2 + |k(t,y)|^2\left|\frac{1}{|t-x|^\zeta}-\frac{1}{|t-y|^\zeta}\right|^2\right) \sigma^{-1}(t)\,dt \\ &\le \text{const }|x-y|^{2\lambda}\int_{-1}^{1}\frac{\sigma^{-1}(t)\,dt}{|t-x|^{2\zeta}} + \text{const }|x-y|^{2\lambda} \\ &\le \text{const }|x-y|^{2\lambda}. \end{aligned}$$

Hence, all functions in $\{Ku : \|u\|_\sigma \le 1\}$ uniformly satisfy a Hölder condition with the exponent λ. It remains to show the uniform boundedness of these functions. Using Lemma 4.11, we get

$$\begin{aligned} |(Ku)(x)| &\le \frac{1}{\pi}\int_{-1}^{1}\left|\frac{w(t)-w(x)}{t-x}\right| |u(t)|\,dt \\ &\le \text{const}\left(\int_{-1}^{1}\frac{\sigma^{-1}(t)\,dt}{|t-x|^{2(1-\eta)}}\right)^{\frac{1}{2}} \|u\|_\sigma \\ &\le \text{const }(1-x)^{-\alpha^+/2}(1+x)^{-\beta^+/2}\,\|u\|_\sigma . \end{aligned}$$

In particular, $\|u\|_\sigma \le 1$ implies $|(Ku)(0)| \le$ const , which together with the uniform Hölder condition results in $\|Ku\|_\infty \le$ const . Thus, we have $K \in \mathcal{L}(\mathbf{L}^2_\sigma, \mathbf{C}^{0,\lambda})$ for some $\lambda > 0$, and the assertion follows if we note that the embedding $\mathbf{C}^{0,\lambda} \subset \mathbf{C}^{0,\lambda'}$ is compact for $\lambda > \lambda'$. ■

Using Lemma 4.13 and Corollary 4.5, we can now estimate the third summand:

$$\|\widetilde{L^\sigma_n} w^{-1} K P^\sigma_n\|_{\mathcal{L}(\mathbf{L}^2_\sigma)} \le \|\widetilde{L^\sigma_n}\|_{\mathcal{L}(\mathbf{C}^{0,\lambda},\mathbf{L}^2_\sigma)}\,\|w^{-1}\|_{\mathbf{C}^{0,\lambda}}\,\|K\|_{\mathcal{L}(\mathbf{L}^2_\sigma,\mathbf{C}^{0,\lambda})} \le \text{const}\,.$$

Thus, we have proved the uniform boundedness of $\widetilde{L^\sigma_n} S P^\sigma_n$. If we note the obvious identity $\widetilde{L^\sigma_n} b S P^\sigma_n = \widetilde{L^\sigma_n} b P^\sigma_n\,\widetilde{L^\sigma_n} S P^\sigma_n$, we can summarize the results of this subsection as follows:

Theorem 4.14 *Let a, b be Riemann integrable on $[-1,1]$. Then the operators $A_n = \widetilde{L^\sigma_n}(aI+bS)P^\sigma_n$ converge strongly to $aI+bS$ on $\mathbf{L}^2_\sigma$.*

4.2 Strong convergence of A_n^*

In all what follows we identify the dual space of $\mathbf{L}^2_\sigma$ with $\mathbf{L}^2_\sigma$ itself and consider A_n^* as an element of $\mathcal{L}(\mathbf{L}^2_\sigma)$. As an auxiliary relation, we deduce a formula for the Fourier coefficients of $\widetilde{L_n^\sigma} f$: We have $\widetilde{L_n^\sigma} f = \sum_{k=0}^{n-1} \alpha_k \widetilde{u}_k$, where, because of the exactness of the Gaussian quadrature rule,

$$\begin{aligned} \alpha_k &= \sum_{s=0}^{n-1} \alpha_s \overbrace{\sum_{j=1}^{n} \widetilde{A_{jn}^\sigma} \widetilde{u}_k(x_{jn}^v) \widetilde{u}_s(x_{jn}^v)}^{=\delta_{ks}} = \sum_{j=1}^{n} \widetilde{A_{jn}^\sigma} \widetilde{u}_k(x_{jn}^v) \sum_{s=0}^{n-1} \alpha_s \widetilde{u}_s(x_{jn}) \\ &= \sum_{j=1}^{n} \widetilde{A_{jn}^\sigma} \widetilde{u}_k(x_{jn}^v) f(x_{jn}^v), \end{aligned} \tag{4.2}$$

where $\widetilde{A_{jn}^\sigma} := w_{v,\sigma^{-1}}^{-2}(x_{jn}^v) A_{jn}^v$ and δ_{ks} denotes the Kronecker symbol.

Now we compute $(\widetilde{L_n^\sigma} a P_n^\sigma)^*$ with Riemann integrable a. For $u = \sum_{k=0}^\infty u_k \widetilde{u}_k$, $v = \sum_{k=0}^\infty v_k \widetilde{u}_k$ we have due to (4.2)

$$\begin{aligned} \langle \widetilde{L_n^\sigma} a P_n^\sigma u, v \rangle_\sigma &= \sum_{k=0}^{n-1} \overline{v_k} \left(\sum_{j=1}^{n} \widetilde{A_{jn}^\sigma} a(x_{jn}^v) \sum_{s=0}^{n-1} u_s \widetilde{u}_s(x_{jn}^v)\, \widetilde{u}_k(x_{jn}^v) \right) \\ &= \sum_{s=0}^{n-1} u_s \overline{\left(\sum_{j=1}^{n} \widetilde{A_{jn}^\sigma} \overline{a}(x_{jn}^v) \sum_{k=0}^{n-1} v_k \widetilde{u}_k(x_{jn}^v)\, \widetilde{u}_s(x_{jn}^v) \right)} = \langle u, \widetilde{L_n^\sigma} \overline{a} P_n^\sigma v \rangle_\sigma, \end{aligned}$$

that means

$$(\widetilde{L_n^\sigma} a P_n^\sigma)^* = \widetilde{L_n^\sigma} \overline{a} P_n^\sigma,$$

which is strongly convergent due to Proposition 4.6.

Since we have $(\widetilde{L_n^\sigma} b S P_n^\sigma)^* = (\widetilde{L_n^\sigma} S P_n^\sigma)^* (\widetilde{L_n^\sigma} b P_n^\sigma)^*$, we can now restrict ourselves to investigating $(\widetilde{L_n^\sigma} S P_n^\sigma)^*$. This will be done again by using a three-term decomposition according to (4.1). The multiplication operator was already considered. To deal with $(\widetilde{a} I + iS) w I$ (we can obviously neglect the factor w^{-1}), we note that in Proposition 1.1 there is always $\kappa \in \{-1, 0, 1\}$, which implies that

$$q_{n-\kappa} := (\widetilde{a} I + iS) w P_n^\sigma u$$

is always a polynomial of degree at most n. If $u, v \in \mathbf{L}^2_\sigma$ and $\widetilde{A_{jn}^\sigma}$ are as above, we can write

$$\begin{aligned} \langle \widetilde{L_n^\sigma} (\widetilde{a} I + iS) w P_n^\sigma u, v \rangle_\sigma &= \sum_{j=1}^{n} \widetilde{A_{jn}^\sigma} q_{n-\kappa}(x_{jn}^v) \overline{(P_n^\sigma v)(x_{jn}^v)} \\ &= \langle w_{v,\sigma^{-1}} q_{n-\kappa}, \widetilde{L_n^\sigma} w_{v,\sigma^{-1}}^{-1} P_n^\sigma v \rangle_\sigma = \langle w_{v,\sigma^{-1}} (\widetilde{a} I + iS) w P_n^\sigma u, \widetilde{L_n^\sigma} w_{v,\sigma^{-1}}^{-1} P_n^\sigma v \rangle_\sigma \\ &= \langle u, P_n^\sigma w (\widetilde{a} I + iS)^* w_{v,\sigma^{-1}} \widetilde{L_n^\sigma} w_{v,\sigma^{-1}}^{-1} P_n^\sigma v \rangle_\sigma. \end{aligned}$$

(Note that $\widetilde{a} I + iS \in \mathcal{L}(\mathbf{L}^2_\sigma)$ and $w_{v,\sigma^{-1}} \widetilde{L_n^\sigma} f \in \mathbf{L}^2_\sigma$ for all f.) Hence, we have

$$(\widetilde{L_n^\sigma} (\widetilde{a} I + iS) w P_n^\sigma)^* = P_n^\sigma w (\widetilde{a} I + iS)^* w_{v,\sigma^{-1}} \widetilde{L_n^\sigma} w_{v,\sigma^{-1}}^{-1} P_n^\sigma.$$

Lemma 4.15 (Cf. [38], Th. 3.1; [9], Equ. (6.2.6); [37], Th. 2.58) *Let $u \in \mathbf{L}^2$ be a Jacobi weight satisfying*

$$\frac{u}{\sqrt{v\varphi}}, \quad \frac{\sqrt{v\varphi}}{u} \in \mathbf{L}^2.$$

If f is a function with fu, $f'\varphi u \in \mathbf{L}^2$, then the following estimation holds:

$$\|u(L_n^v f - f)\|_{\mathbf{L}^2} \le \frac{\text{const}}{n^{\frac{1}{2}}} \|f'\varphi u\|_{\mathbf{L}^2}.$$

Lemma 4.16 *Let $\gamma, \delta \in (0,1)$ be chosen such that $\gamma > \alpha - \frac{1}{2}$, $\delta > \beta - \frac{1}{2}$. Then the operators $P_n^\sigma w(\widetilde{a}I + iS)^* w_{v,\sigma^{-1}} \widetilde{L_n^\sigma} w_{v,\sigma^{-1}}^{-1} P_n^\sigma$ converge strongly in $\mathbf{L}_\sigma^2$.*

Proof. The uniform boundedness is trivial in view of Proposition 4.10, thus we only have to show the convergence on span $\{\widetilde{u}_m\}_{m=0}^\infty$. Since $P_n^\sigma \to I$ and $(\widetilde{a}I + iS)^* \in \mathcal{L}(\mathbf{L}_\sigma^2)$, we only need to consider the term $w_{v,\sigma^{-1}} \widetilde{L_n^\sigma} w_{v,\sigma^{-1}}^{-1} I$. Let $m \ge n$. We can write

$$\left\| w_{v,\sigma^{-1}} \widetilde{L_n^\sigma} w_{v,\sigma^{-1}}^{-1} \widetilde{u}_m - \widetilde{u}_m \right\|_\sigma = \|\widetilde{L_n^\sigma} p_m^v - p_m^v\|_v$$

$$= \|v^{1/2} w_{v,\sigma^{-1}} (L_n^v w_{v,\sigma^{-1}}^{-1} p_m^v - w_{v,\sigma^{-1}}^{-1} p_m^v)\|_{\mathbf{L}^2}.$$

We now apply Lemma 4.15 with $f = w_{v,\sigma^{-1}}^{-1} p_m^v$, $u = v^{1/2} w_{v,\sigma^{-1}} = v\sigma^{-1/2}$. Because of γ, $\delta > 0$ we have $u \in \mathbf{L}^2$. Further,

$$\frac{u}{\sqrt{v\varphi}} = (1-x)^{(\gamma-\alpha)/2-1/4}(1+x)^{(\delta-\beta)/2-1/4},$$

hence the other conditions on u and v in Lemma 4.15 are satisfied if and only if the relations $-1 < \gamma - \alpha - \frac{1}{2}$, $\delta - \beta - \frac{1}{2} < 1$ hold, or equivalently,

$$\alpha - \frac{1}{2} < \gamma < \alpha + \frac{3}{2}, \qquad \beta - \frac{1}{2} < \delta < \beta + \frac{3}{2}. \tag{4.3}$$

Now let us estimate $(w_{v,\sigma^{-1}}^{-1} p_m^v)'\varphi u$. We have

$$(w_{v,\sigma^{-1}}^{-1})' = O\left((1-x)^{\frac{\alpha-\gamma}{2}-1}(1+x)^{\frac{\beta-\delta}{2}-1}\right).$$

(Note that we excluded the cases $\alpha = \gamma$ and $\beta = \delta$.) Hence,

$$(w_{v,\sigma^{-1}}^{-1} p_m^v)'\varphi u = O\left((1-x)^{\frac{\gamma-1}{2}}(1+x)^{\frac{\delta-1}{2}}\right),$$

which is obviously in $\mathbf{L}^2$. Thus, the assertion is proved if the exponents γ, δ of v satisfy (4.3). If, for instance, $\gamma \ge \alpha + \frac{3}{2}$, we choose a number $\widetilde{\gamma}$, $0 < \widetilde{\gamma} < \gamma$, such that

$$\alpha - \frac{1}{2} < 2\widetilde{\gamma} - \gamma < \alpha + \frac{3}{2}, \qquad 2\widetilde{\gamma} - \gamma > -1,$$

and take $\widetilde{u} := \widetilde{v}\sigma^{-1/2}$, where $\widetilde{v}(x) = (1-x)^{\widetilde{\gamma}}(1+x)^\delta$. Then the conditions on the weights in Lemma 4.15 are again satisfied with $\widetilde{u}$ instead of u, and we can make use of $\|u(L_n^v f - f)\|_{\mathbf{L}^2} \le \text{const}\, \|\widetilde{u}(L_n^v f - f)\|_{\mathbf{L}^2} \to 0$. ■

Due to Lemma 4.13, we have $K := wS - SwI \in \mathcal{K}(\mathbf{L}^2_\sigma, \mathbf{C}^{0,\lambda})$ for some $\lambda > 0$. Furthermore, $(P^\sigma_n)^* = P^\sigma_n \to I$ in $\mathbf{L}^2_\sigma$ and $\widetilde{L^\sigma_n} \to E$ (cf. Corollary 4.5), where E denotes the continuous embedding of $\mathbf{C}^{0,\lambda}$ into $\mathbf{L}^2_\sigma$. Thus, if we write again K instead of EK, Lemma 3.10 gives

$$\|\widetilde{L^\sigma_n} K P^\sigma_n - K\|_{\mathcal{L}(\mathbf{L}^2_\sigma)} \longrightarrow 0 \qquad (n \to \infty),$$

and hence

$$\|(\widetilde{L^\sigma_n} K P^\sigma_n)^* - K^*\|_{\mathcal{L}(\mathbf{L}^2_\sigma)} \longrightarrow 0 \qquad (n \to \infty).$$

Thus, we have proved the following theorem:

Theorem 4.17 *If a, b are Riemann integrable, $\gamma > \max\{\alpha - \frac{1}{2}, 0\}$, $\delta > \max\{\beta - \frac{1}{2}, 0\}$, then the sequence $(\widetilde{L^\sigma_n}(aI + bS)P^\sigma_n)^*$ is strongly convergent.*

4.3 The cases $\gamma = \alpha$, $\delta = \beta$

If one of the numbers $\alpha_0 := \frac{\gamma - \alpha}{2}$, $\beta_0 := \frac{\delta - \beta}{2}$ is zero, the decomposition (4.1), on which our proof of the uniform boundedness of $\widetilde{L^\sigma_n} S P^\sigma_n$ was based, is not possible. We cannot choose $\widetilde{b} \equiv 1$ in Proposition 1.1 since in this case $\alpha_0 - g(1)$ (or $\beta_0 + g(-1)$, respectively) cannot be an integer. We can, however, to some extent overcome the difficulties connected with this fact if we estimate the whole term $\widetilde{L^\sigma_n} bSP^\sigma_n$ instead of considering $\widetilde{L^\sigma_n} S P^\sigma_n$ separately.

Assume for instance $\alpha_0 = 0$, $\beta_0 \neq 0$. In this case we can proceed as follows. Let b be a function satisfying the following conditions:

$$\begin{cases} b \in \mathbf{C}^{0,\eta} \text{ for some } \eta > \frac{1+\max\{\alpha,\beta,0\}}{2} \\ b(1) = 0 \\ b(-1) \neq 0. \end{cases} \tag{4.4}$$

(If $\alpha_0 = \beta_0 = 0$, we would also require $b(-1) = 0$). For the following argument we can assume without loss of generality that b is real-valued and $b(-1) < 0$. Now we choose some $\widetilde{a} \in \mathbf{C}^1$, $\widetilde{a}(1) = 1$ (that is, $g(1) = 0$, cf. the notations of Prop. 1.1) such that the operator $(\widetilde{a}I + iSbI)w_{v,\sigma^{-1}}wI$ possesses the mapping properties described in Proposition 1.1 with $\kappa \geq -1$. If $\beta_0 > 0$, we can choose $\widetilde{a}$ such that $g(-1) = 1 - \beta_0$, whence $\kappa = -1$, if $\beta_0 < 0$ we choose $g(-1) = -\beta_0$ and have $\kappa = 0$. (If we had $\beta_0 = 0$, we would simply take $\widetilde{a} \equiv 1$, which means $\kappa = 0$.)

Instead of (4.2) we now use the decomposition

$$bS = SbI + K = i\widetilde{a}I - iw^{-1}(\widetilde{a}I + iSbI)wI + w^{-1}(wS - SwI)bI + K,$$

where $K = bS - SbI \in \mathcal{K}(\mathbf{L}^2_\sigma, \mathbf{C}^{0,\mu})$ for some $\mu > 0$ (compare Lemma 4.13). This enables us to show in the same way as before that $A_n = \widetilde{L^\sigma_n} bSP^\sigma_n$ and A^*_n converge strongly if b satisfies (4.4). In particular, we have $\|\widetilde{L^\sigma_n}(1-x)^\eta SP^\sigma_n\| \leq \text{const}$ if $\eta > \frac{1+\max\{\alpha,\beta,0\}}{2}$.

Proposition 4.18 *Let $\alpha_0 = 0$, $\beta_0 \neq 0$. Let $b \in \mathbf{PC}$ and*

$$b(x) = o((1-x)^\eta) \qquad (x \to 1), \qquad \eta > \frac{1+\max\{\alpha,\beta,0\}}{2}.$$

Then $\|\widetilde{L_n^\sigma} bSP_n^\sigma\| \leq \mathrm{const}$. (If $\beta_0 = 0$, we would require an analogous behaviour near the point -1.)

Proof. Let χ be Riemann integrable, and let $\widetilde{b} \in \Pi_{1,0}$, which denotes the set of all polynomials vanishing in 1. Then $\widetilde{L_n^\sigma}\chi\widetilde{b}SP_n^\sigma = \widetilde{L_n^\sigma}\chi(1+x)^k P_n^\sigma \widetilde{L_n^\sigma}\widetilde{b}_1 SP_n^\sigma$ with some nonnegative integer k, where $\widetilde{b}_1$ satisfies (4.4). Hence, the uniform boundedness of $\widetilde{L_n^\sigma} bSP_n^\sigma$ continues to hold if b is the product of a polynomial from $\Pi_{1,0}$ with a Riemann integrable function.

Now let b be as in the hypothesis with a finite number of jumps. Then $(1-x)^{-\eta}b$ is piecewise continuous and can therefore be approximated uniformly by a piecewise polynomial $\widetilde{b} = \sum_{j=1}^m \chi_j \widetilde{b}_j$, where $\widetilde{b}_j \in \Pi_{1,0}$, χ_j is the characteristic function of $[x_j, x_{j+1}]$ and $-1 = x_1 < x_2 < \ldots < x_{m+1} = 1$. (Note that $\Pi_{1,0}$ is dense in $\mathbf{C}[x_j, x_{j+1}]$ for all $j = 1, \ldots, m-1$ due to the Stone-Weierstraß theorem.) Then we have

$$\left\|\widetilde{L_n^\sigma}\left((1-x)^\eta \widetilde{b} - b\right) SP_n^\sigma\right\| \leq \left\|\widetilde{L_n^\sigma}\left(\widetilde{b} - (1-x)^{-\eta} b\right) P_n^\sigma\right\| \left\|\widetilde{L_n^\sigma}(1-x)^\eta SP_n^\sigma\right\|$$

$$\leq \mathrm{const} \left\|\widetilde{b} - (1-x)^{-\eta} b\right\|_\infty,$$

which can be made as small as desired by the choice of $\widetilde{b}$. If we note the fact that the set of all sequences from $\mathcal{E}$ which, together with their adjoint operators, are strongly convergent is a closed subalgebra of $\mathcal{E}$ (compare the proof of Lemma 3.9), we get the assertion for b with a finite number of jumps. If $b \in \mathbf{PC}$ is arbitrary, $b(x) = o((1-x)^\eta)$, we approximate $(1-x)^{-\eta}b$ uniformly by a function $\widetilde{b}$ with the same properties and only finitely many jumps and repeat the same arguments as above to get the assertion for the coefficient b. ∎

4.4 Strong convergence of $\widetilde{A_n}$

Though until now we have considered an interpolation with respect to the nodes x_{jn}^v, where v is an arbitrary Jacobi weight satisfying (1.4), we will in the following confine ourselves to the Chebyshev weight of second kind $v(x) = \varphi(x)$, since some of the following proofs make use of the special form of the corresponding orthogonal polynomials $U_n = p_n^\varphi$. In the sequel we will therefore briefly write w_σ instead of $w_{v,\sigma}$. First we consider the case of a multiplication operator $A = aI$ with a Riemann integrable function a. The following lemma together with Proposition 4.6 shows the convergence of $\widetilde{A_n}$ in this case.

Lemma 4.19 *We have $W_n^\sigma \widetilde{L_n^\sigma} a W_n^\sigma = \widetilde{L_n^\sigma} a P_n^\sigma$ for all n.*

Proof. It is sufficient to show the identity on span $\{\widetilde{u}_m\}$. For $n > m$ we have (compare relation (4.2))

$$W_n^\sigma \widetilde{L_n^\sigma} a W_n^\sigma \widetilde{u}_m = W_n^\sigma \widetilde{L_n^\sigma} a \widetilde{u}_{n-1-m}$$

$$= W_n^\sigma \sum_{k=0}^{n-1} \left(\sum_{j=1}^{n} \widetilde{A_{jn}^\sigma} \widetilde{u}_k(x_{jn}^\varphi) \widetilde{u}_{n-1-m}(x_{jn}^\varphi) a(x_{jn}^\varphi) \right) \widetilde{u}_k$$

$$= \sum_{k=0}^{n-1} \left(\sum_{j=1}^{n} \widetilde{A_{jn}^\sigma} \widetilde{u}_{n-1-k}(x_{jn}^\varphi) \widetilde{u}_{n-1-m}(x_{jn}^\varphi) a(x_{jn}^\varphi) \right) \widetilde{u}_k =: \sum_{k=0}^{n-1} \beta_{km} \widetilde{u}_k.$$

If we remember that $x_{jn}^\varphi = \cos\frac{j\pi}{n+1}$ $(j = 1, \dots, n)$ and $A_{jn}^\varphi = \pi \frac{1-(x_{jn}^\varphi)^2}{n+1}$, we get

$$\beta_{km} = \frac{2}{n+1} \sum_{j=1}^{n} \sin\frac{(n-k)j\pi}{n+1} \sin\frac{(n-m)j\pi}{n+1} a(x_{jn}^\varphi)$$

$$= \frac{2}{n+1} \sum_{j=1}^{n} \sin\frac{(k+1)j\pi}{n+1} \sin\frac{(m+1)j\pi}{n+1} a(x_{jn}^\varphi)$$

$$= \beta_{n-1-k,n-1-m},$$

which means $W_n^\sigma \widetilde{L_n^\sigma} a W_n^\sigma \widetilde{u}_m = \widetilde{L_n^\sigma} a P_n^\sigma \widetilde{u}_m$. ■

For $x \in [-1, 1]$ let $e(x) = x$. Now we introduce the operator $V := eI - iw_{\sigma^{-1}} S w_\sigma I$, that is, $(Vu)(x) = xu(x) - iw_{\sigma^{-1}}(x)(Sw_\sigma u)(x)$ for $u \in \mathbf{L}_\sigma^2$. The following lemma is a generalization of a result from [48] (cf. also [44, Th. 4.123]), which is formulated there for the case $\alpha = \beta = 0$.

Lemma 4.20 ([48], see also [44], Theorem 4.123) *V is a left shift operator with respect to the system $\{\widetilde{u}_n\}_{n=0}^\infty$, more precisely, the relation*

$$Vu = \sum_{k=0}^{\infty} \langle u, \widetilde{u}_k \rangle_\sigma \widetilde{u}_{k+1}$$

holds for $u \in \mathbf{L}_\sigma^2$. The adjoint operator, which satisfies

$$V^* u = \sum_{k=0}^{\infty} \langle u, \widetilde{u}_{k+1} \rangle_\sigma \widetilde{u}_k,$$

is given by

$$V^* = eI + iw_{\sigma^{-1}} S w_\sigma I.$$

Proof. For the shift property of V compare the proof of [44, Theorem 4.123]. To verify the representation of V^*, note that $\sigma^{1/2} I : \mathbf{L}_\sigma^2 \to \mathbf{L}^2$ is an isometric isomorphism and $S^* = S$ in $\mathbf{L}^2$. ■

Lemma 4.21 ([36], Lemma 3.10, cf. also [42], §5) *Assume that*

$$b \in \mathbf{C}^{p,\eta}[-1,1] \quad (0<\eta\le 1) \qquad \text{and} \qquad b^{(j)}(\pm 1)=0 \quad (j=0,\dots,p).$$

Let further $v=v^{-\gamma,-\delta}$ *and* $\lambda:=\eta-\max\{\gamma,\delta,0\}>0$. *Then*

$$bv \in \mathbf{C}^{p,\lambda}[-1,1] \qquad \text{and} \qquad (bv)^{(j)}(\pm 1)=0 \quad (j=0,\dots,p).$$

We introduce the notations

$$\mathbf{PC}_0 := \{b\in \mathbf{PC}[-1,1] : b(\pm 1)=0\},$$

$$\mathbf{C}_0 := \{b\in \mathbf{C}[-1,1] : b(\pm 1)=0\}$$

and

$$\mathbf{C}^{1,\eta}_{0,0} := \left\{b\in \mathbf{C}^{1,\eta}[-1,1] : b(\pm 1)=b'(\pm 1)=0\right\}.$$

Proposition 4.22 *Let* $b\in\mathbf{PC}_0$. *Then the sequence* $W_n^\sigma \widetilde{L_n^\sigma} bSW_n^\sigma$ *converges strongly to* $-bw_\sigma^{-1}Sw_\sigma I$.

Proof. First let b be of the form $b=\chi b_1$, where $b_1\in\mathbf{C}^{1,\eta}_{0,0}$ with $\eta>\max\{\frac14+\frac{\alpha}{2},\frac14+\frac{\beta}{2},0\}$ and $\chi\in\mathbf{R}$, which denotes the space of Riemann integrable functions. We use the following three-term decomposition of b_1S (cf. Lemma 4.20):

$$b_1S = b_1w_\sigma^{-1}Sw_\sigma I + K_1 + K_2 = ib_1\varphi^{-1}(eI-V^*)+K_1+K_2, \tag{4.5}$$

where $K_1=b_1S-Sb_1I$ and $K_2=(Sb_1w_\sigma^{-1}I-b_1w_\sigma^{-1}S)w_\sigma I$. Using Lemma 4.19, we can manage the first two summands of $\chi b_1 S$ (note that $b_1\varphi^{-1}$ is bounded) according to (4.5):

$$iW_n^\sigma\widetilde{L_n^\sigma}\chi b_1\varphi^{-1}eW_n^\sigma = i\widetilde{L_n^\sigma}\chi b_1\varphi^{-1}eP_n^\sigma \to i\chi b_1\varphi^{-1}eI$$

and

$$-iW_n^\sigma\widetilde{L_n^\sigma}\chi b_1\varphi^{-1}V^*W_n^\sigma = -iW_n^\sigma\widetilde{L_n^\sigma}\chi b_1\varphi^{-1}W_n^\sigma\cdot W_n^\sigma V^* W_n^\sigma$$

$$= -i\widetilde{L_n^\sigma}\chi b_1\varphi^{-1}P_n^\sigma\cdot P_n^\sigma V \to -i\chi b_1\varphi^{-1}V.$$

The multiplication operator $w_\sigma I$ is an isometric isomorphism from $\mathbf{L}^2_\sigma$ onto $\mathbf{L}^2_{\varphi^{-1}}$. Since $b_1\in\mathbf{C}^{1,\eta}_{0,0}$, we get $b_1\varphi^{-1}\in\mathbf{C}^1$ from Lemma 4.21, and Lemma 4.13 gives us $Sb_1w_\sigma^{-1}I-b_1w_\sigma^{-1}S\in\mathcal{K}(\mathbf{L}^2_{\varphi^{-1}},\mathbf{C}^{0,\lambda})$ as well as $K_1\in\mathcal{K}(\mathbf{L}^2_\sigma,\mathbf{C}^{0,\lambda})$ with some $\lambda>0$. Hence, $K:=\chi(K_1+K_2)\in\mathcal{K}(\mathbf{L}^2_\sigma,\mathbf{R})$. Since W_n^σ converges weakly to 0, we have $KW_n^\sigma\to 0$ strongly in $\mathcal{L}(\mathbf{L}^2_\sigma,\mathbf{R})$ by virtue of Lemma 3.10. Moreover, we have

$$\left\|W_n^\sigma\widetilde{L_n^\sigma}\right\|_{\mathcal{L}(\mathbf{R},\mathbf{L}^2_\sigma)} \le \|W_n^\sigma\|_{\mathcal{L}(\mathbf{L}^2_\sigma)}\,\|\widetilde{L_n^\sigma}\|_{\mathcal{L}(\mathbf{R},\mathbf{L}^2_\sigma)} \le \text{const},$$

which results in $W_n^\sigma\widetilde{L_n^\sigma}KW_n^\sigma\to 0$ (strongly). Thus, we can conclude

$$W_n^\sigma\widetilde{L_n^\sigma}\chi b_1SW_n^\sigma \to i\chi b_1\varphi^{-1}(eI-V) = -\chi b_1w_\sigma^{-1}Sw_\sigma I \tag{4.6}$$

if $b \in \mathbf{C}_{0,0}^{1,\eta}, \chi \in \mathbf{R}$. Now we consider the general case $b \in \mathbf{PC}_0$. Without loss of generality we can restrict ourselves to investigating functions with a finite number of jumps, that is, we can write $b = \sum_{j=1}^m \chi_j b_j$, where $b_j \in \mathbf{C}_0$, $\chi_j = \chi_{[x_j,x_{j+1}]}$ is the characteristic function of the subinterval $[x_j, x_{j+1}]$, and $-1 = x_1 < x_2 < \ldots < x_m < x_{m+1} = 1$ is an arbitrary partition of $[-1, 1]$. Now we approximate b by piecewise $\mathbf{C}_{0,0}^{1,\eta}$-functions: Let $\varepsilon > 0$ and choose functions $\widetilde{b}_j \in \mathbf{C}_{0,0}^{1,\eta}$, $j = 1, \ldots, m$, such that

$$|b_j(x) - \widetilde{b}_j(x)| < \varepsilon, \qquad x \in [x_j, x_{j+1}].$$

If we define $\widetilde{b} := \sum_{j=1}^m \chi_j \widetilde{b}_j$, we obviously have $\|b - \widetilde{b}\|_\infty < \varepsilon$. Since

$$\|\{\widetilde{L_n^\sigma}(b - \widetilde{b})SP_n^\sigma\}\|_{\mathcal{A}} \le \|\widetilde{L_n^\sigma}(b - \widetilde{b})P_n^\sigma\| \, \|\widetilde{L_n^\sigma}SP_n^\sigma\| \le \text{const } \|b - \widetilde{b}\|_\infty < \varepsilon \tag{4.7}$$

and since $\mathcal{A}$ is a closed subalgebra of $\mathcal{E}$, we can conclude from (4.6) together with Theorems 4.14 and 4.17 and Proposition 4.25 of the following subsection that $W_n^\sigma \widetilde{L_n^\sigma} bSW_n^\sigma$ is strongly convergent. The continuity of the homomorphism $\mathcal{W}_1$ and the relation

$$\|(b - \widetilde{b})w_\sigma^{-1}Sw_\sigma I\|_{\mathcal{L}(\mathbf{L}_\sigma^2)} \le \text{const } \|b - \widetilde{b}\|_\infty$$

imply that $\mathcal{W}_1\{\widetilde{L_n^\sigma}bSP_n^\sigma\} = -bw_\sigma^{-1}Sw_\sigma I$ for $b \in \mathbf{PC}_0$ with finitely many jumps. An arbitrary $\mathbf{PC}_0$-function can be approximated uniformly by such functions, and we can repeat the same arguments as above to get the assertion. ■

Remark 4.23 *In the case $\alpha = \frac{1}{2}$ (or $\beta = \frac{1}{2}$) we additionaly require that b satisfies the hypotheses of Proposition 4.18. If b has only finitely many jumps, we can approximate $(1 - x)^{-\eta}b$ by a piecewise $\mathbf{C}_{0,0}^{1,\eta}$-function $\widetilde{b}$, and instead of (4.7), we use the estimation*

$$\left\|\widetilde{L_n^\sigma}\left((1 - x)^\eta \widetilde{b} - b\right) SP_n^\sigma\right\| \le \text{const } \left\|\widetilde{b} - (1 - x)^{-\eta}b\right\|_\infty$$

(compare the proof of Proposition 4.18) and the fact that the assertion of Proposition 4.22 is true for the coefficient $(1-x)^\eta\widetilde{b}$. The transition to an arbitrary number of jumps is analogous.

Remark 4.24 (cf. [20], Theorem IX.4.1) *The operator $\widetilde{A} = aI - bw_\sigma^{-1}Sw_\sigma I$ is invertible in $\mathbf{L}_\sigma^2$ if and only if $aI - bS$ in invertible in $\mathbf{L}_{\varphi^{-1}}^2$. If $a \in \mathbf{PC}$, $b \in \mathbf{PC}_0$, this is equivalent to the invertibility of $A = aI + bS$ in $\mathbf{L}_\sigma^2$ (compare Proposition 5.14).*

4.5 Strong convergence of $\widetilde{A_n}^*$

Proposition 4.25 *If a is Riemann integrable and $b \in \mathbf{PC}_0$, then the operator sequence $\widetilde{A_n}^* = (W_n^\sigma \widetilde{L_n^\sigma}(aI + bS)W_n^\sigma)^*$ is strongly convergent.*

Proof. For the multiplication operator we have

$$(W_n^\sigma \widetilde{L_n^\sigma} a W_n^\sigma)^* = (\widetilde{L_n^\sigma} a P_n^\sigma)^* = \widetilde{L_n^\sigma}\overline{a}P_n^\sigma \longrightarrow \overline{a}I$$

by Lemma 4.19.

First we assume $b = \chi b_1$, where $\chi \in \mathbf{R}$ and $b_1 \in \mathbf{C}^{1,\eta}_{0,0}$. The investigation of $\chi b_1 S$ will again be based on (4.5). The expression $\chi b_1 \varphi^{-1} eI$ is already covered by the preceding arguments. Further we have

$$(W^\sigma_n \widetilde{L^\sigma_n} \chi b_1 \varphi^{-1} V^* W^\sigma_n)^* = (W^\sigma_n \widetilde{L^\sigma_n} \chi b_1 \varphi^{-1} W^\sigma_n \cdot W^\sigma_n V^* W^\sigma_n)^*$$
$$= W^\sigma_n V W^\sigma_n \widetilde{L^\sigma_n} \overline{\chi b_1} \varphi^{-1} P^\sigma_n = V^* P^\sigma_n \widetilde{L^\sigma_n} \overline{\chi b_1} \varphi^{-1} P^\sigma_n \to V^* \overline{\chi b_1} \varphi^{-1} I.$$

Since $K := \chi(K_1 + K_2) \in \mathcal{K}(\mathbf{L}^2_\sigma, \mathbf{R})$ and hence $\|\widetilde{L^\sigma_n} K - K\|_{\mathcal{L}(\mathbf{L}^2_\sigma)} \to 0$, we have the relation $\|(\widetilde{L^\sigma_n} K)^* - K^*\| \to 0$. Thus, we can write

$$(W^\sigma_n \widetilde{L^\sigma_n} K W^\sigma_n)^* = W^\sigma_n \left((\widetilde{L^\sigma_n} K)^* - K^* \right) W^\sigma_n + W^\sigma_n K^* W^\sigma_n,$$

the first summand uniformly and the second strongly converging to 0 (compare Lemma 3.10). By approximation we can finally get the assertion for arbitrary $b \in \mathbf{PC}_0$ (compare the proof of Proposition 4.22). ■

4.6 The special case $v = \varphi$, $\sigma = \varphi^{-1}$

In the case $v = \varphi$, $\sigma = \varphi^{-1}$, we are in a position to show the strong convergence of $\widetilde{A_n}$ and $\widetilde{A_n}^*$ without the additional assumption that $b(\pm 1) = 0$. This is due to the fact that we have $w_{\sigma^{-1}} = \varphi$, and the mapping properties (2.2) of the operator $S\varphi I$ with respect to the Chebyshev polynomials of second kind U_n enable us to consider the operator sequence $W^\sigma_n \widetilde{L^\sigma_n} S W^\sigma_n$ separately. The following proposition shows that the convergence of this operator and its adjoint can be reduced to the Theorems 4.14 and 4.17.

Proposition 4.26 *Let $v = \varphi$, $\sigma = \varphi^{-1}$. Then we have the identity*

$$W^\sigma_n \widetilde{L^\sigma_n} S W^\sigma_n \equiv -\widetilde{L^\sigma_n} S P^\sigma_n.$$

Proof. For $k, l = 0, \ldots, n-1$ we have

$$\langle W^\sigma_n \widetilde{L^\sigma_n} S W^\sigma_n \widetilde{u}_k, \widetilde{u}_l \rangle_\sigma = \langle \widetilde{L^\sigma_n} S \varphi U_{n-1-k}, \widetilde{u}_{n-1-l} \rangle_\sigma$$
$$= i \langle \widetilde{L^\sigma_n} T_{n-k}, \widetilde{u}_{n-1-l} \rangle_\sigma = i \sum_{j=1}^{n} \widetilde{A^\sigma_{jn}} T_{n-k}(x^\varphi_{jn}) \widetilde{u}_{n-1-l}(x^\varphi_{jn})$$
$$= i \frac{2}{n+1} \sum_{j=1}^{n} \cos \frac{(n-k)j\pi}{n+1} \sin \frac{(n-l)j\pi}{n+1}$$
$$= -i \frac{2}{n+1} \sum_{j=1}^{n} \cos \frac{(k+1)j\pi}{n+1} \sin \frac{(l+1)j\pi}{n+1}$$
$$= -i \sum_{j=1}^{n} \widetilde{A^\sigma_{jn}} T_{k+1}(x^\varphi_{jn}) \widetilde{u}_l(x^\varphi_{jn}) = -\langle \widetilde{L^\sigma_n} S P^\sigma_n \widetilde{u}_k, \widetilde{u}_l \rangle_\sigma. \quad ■$$

Corollary 4.27 *In the case $v = \varphi$, $\sigma = \varphi^{-1}$ we have $\{A_n\} = \{\widetilde{L^\sigma_n}(aI + bS)P^\sigma_n\} \in \mathcal{A}$ for all $a, b \in \mathbf{PC}$, where*

$$\mathrm{s} - \lim_{n\to\infty} A_n = aI + bS \quad \text{and} \quad \mathrm{s} - \lim_{n\to\infty} \widetilde{A_n} = aI - bS.$$

5 Local theory of the collocation method

To prove the stability of the collocation method (2.6), we apply Theorem 3.12. Having shown that the sequence

$$\{A_n\} = \{\widetilde{L_n^\sigma}(aI + bS)P_n^\sigma\}$$

belongs to the algebra $\mathcal{A}$ described in Section 3 and having computed the strong limit $\widetilde{A} = aI - w_\sigma^{-1}bSw_\sigma I$ (cf. Lemma 4.19 and Proposition 4.22), we are left with investigating the invertibility of the coset $\{A_n\} + \mathcal{I} \in \mathcal{A}/\mathcal{I}$, which will be done by the local principle of Gohberg and Krupnik presented in Theorem 3.4. We will again restrict ourselves to the case $v = \varphi$ in the present section.

First of all we choose an appropriate system of localizing classes. For $\tau \in [-1,1]$, let

$$m_\tau := \{f \in \mathbf{C}[-1,1] : 0 \le f(x) \le 1\,, f(x) \equiv 1 \text{ in some neighbourhood of } \tau\}$$

and define

$$M_\tau := \{\{\widetilde{L_n^\sigma} f P_n^\sigma\} + \mathcal{I} : f \in m_\tau\}.$$

Lemma 5.1 (compare [31], Lemma 2.6) $\{M_\tau\}_{\tau\in[-1,1]}$ *is a covering system of localizing classes in* $\mathcal{A}/\mathcal{I}$.

To be able to apply Theorem 3.4, we have to verify that the coset $\{A_n\} + \mathcal{I}$ commutes with all elements of $\bigcup\limits_{\tau\in[-1,1]} M_\tau$. For this end, we will make use of a transformation to the unit circle. It will turn out that there are certain close relations between our collocation method (2.6) on the interval and the collocation method with the Multhopp nodes for singular integral operators on the unit circle considered in [31]. Nevertheless we will demonstrate that it is not possible to simply reduce the question for the stability of the collocation method (2.6) to that one of the method from [31] (see Remark 5.8). Our method requires a self-contained investigation based on Theorem 3.12 and the local principle, Theorem 3.4, and the transformation to the unit circle can merely be used as an auxiliary tool in the local theory. It was for this reason that the author tried to confine the investigation as far as possible to the interval, even if at some places the transition to the unit circle would have made the discussion briefer.

5.1 Associated operators on the unit circle

Let $e_n(t) = t^n$, $n = 0, \pm 1, \pm 2, \ldots$, $t \in \mathbb{T}$. Then $\{e_n\}_{n=-\infty}^{\infty}$ forms an orthonormal basis in $\mathbf{L}^2(\mathbb{T})$, equipped with the usual inner product

$$\langle f, g\rangle = \frac{1}{2\pi}\int_{-\pi}^{\pi} f(e^{is})\overline{g(e^{is})}\,ds.$$

Define the operator $F : \mathbf{L}^2_\sigma \to \mathbf{L}^2(\mathbb{T})$ by

$$Fu = \frac{1}{\sqrt{2}i}\sum_{n=0}^{\infty}\langle u, \widetilde{u}_n\rangle_\sigma (e_{n+1} - e_{-n-1}).$$

Then F is an isometric isomorphism from $\mathbf{L}^2_\sigma$ onto the subspace

$$\mathbf{L}^2_{\text{odd}}(\mathbb{T}) := \operatorname{clos\,span}\{e_n - e_{-n} : n = 1, 2, \ldots\}$$

of all "odd" functions from $\mathbf{L}^2(\mathbb{T})$. Moreover, because of the relation

$$(Fu)(e^{\pm is}) = \pm\sqrt{2}\sum_{n=0}^{\infty}\langle u, \widetilde{u}_n\rangle_\sigma \sin(n+1)s$$

$$= \pm\sqrt{\pi}\sum_{n=0}^{\infty}\langle u, \widetilde{u}_n\rangle_\sigma w_\sigma(\cos s)\,\widetilde{u}_n(\cos s) = \pm\sqrt{\pi}\,w_\sigma(\cos s)\,u(\cos s), \quad 0 < s < \pi,$$

we have

$$(Fu)(t) = \sqrt{\pi}\,\psi(t)\,w_\sigma(\mathsf{Re}\ t)u(\mathsf{Re}\ t),$$

where

$$\psi(t) = \begin{cases} 1 , & \mathsf{Im}\ t > 0, \\ -1 , & \mathsf{Im}\ t < 0, \\ 0 , & t = \pm 1, \end{cases} \tag{5.1}$$

and $\mathsf{Re}\ t$ and $\mathsf{Im}\ t$ denote the real and the imaginary part of a complex number t, respectively. Of course, the preceding equations have to be understood in the sense of "almost everywhere", hence at this place we could give $\psi(\pm 1)$ other values. But, in particular for the application of interpolation operators, it will turn out that we need exactly the above definition of ψ. We furthermore remark that the inverse operator $F^{-1} : \mathbf{L}^2_{\text{odd}}(\mathbb{T}) \to \mathbf{L}^2_\sigma$ can be described by

$$(F^{-1}f)(\cos s) = \frac{1}{\sqrt{\pi}} w_\sigma^{-1}(\cos s) f(e^{is}), \qquad 0 < s < \pi.$$

In the following we will consider the Fourier projections $P_n^{\mathbf{T}} : \mathbf{L}^2(\mathbb{T}) \to \mathbf{L}^2(\mathbb{T})$ given by

$$P_n^{\mathbf{T}} f = \sum_{k=-n-1}^{n} \langle f, e_k\rangle e_k$$

and operators $W_n^{\mathbf{T}} : \mathbf{L}^2(\mathbb{T}) \to \mathbf{L}^2(\mathbb{T})$ defined by

$$W_n^{\mathbf{T}} f = \sum_{k=0}^{n}\langle f, e_{n-k}\rangle e_k + \sum_{k=-n-1}^{-1} \langle f, e_{-n-2-k}\rangle e_k.$$

Furthermore, let $M_n^{\mathbf{T}}$ be the Multhopp interpolation operator that assigns to every Riemann integrable function on $\mathbb{T}$ the (uniquely determined) trigonometric polynomial $M_n^{\mathbf{T}} f = \sum_{k=-n-1}^{n} m_{kn}(f)\, e_k$ such that for the Multhopp nodes

$$t_{jn}^M = \exp\left(\frac{\pi i j}{n+1}\right), \qquad j = -n-1, \ldots, n,$$

the relation

$$(M_n^{\mathbf{T}} f)(t_{jn}^M) = f(t_{jn}^M), \qquad j = -n-1, \ldots, n$$

holds. One can show (see [31]) that

$$m_{kn}(f) = \frac{1}{2n+2} \sum_{j=-n-1}^{n} f(t_{jn}^M)(t_{jn}^M)^{-k}, \qquad k = -n-1, \ldots, n. \tag{5.2}$$

Lemma 5.2 *If $f \in \mathbf{L}^2_{\mathrm{odd}}(\mathbb{T})$, more precisely,*

$$f(e^{is}) = -f(e^{-is}), \qquad 0 < s < \pi, \qquad f(\pm 1) = 0,$$

then we have again $M_n^{\mathbf{T}} f \in \mathbf{L}^2_{\mathrm{odd}}(\mathbb{T})$.

Proof. Suppose that f has the assumed properties. Then (5.2) shows that

$$m_{kn}(f) = -\frac{2i}{2n+2} \sum_{j=1}^{n} f(t_{jn}^M) \sin \frac{jk\pi}{n+1}, \qquad k = -n-1, \ldots, n,$$

whence we get

$$m_{kn}(f) = -m_{-k,n}(f), \qquad k = 1, \ldots, n, \qquad m_{0n}(f) = m_{-n-1,n}(f) = 0,$$

and the lemma is proved. ■

One can further prove (cf. [31, Lemma 2.2] or [44, 7.3(b)]) that for every bounded function $a : \mathbb{T} \to \mathbb{C}$

$$\left\| M_n^{\mathbf{T}} a p_n \right\|_{\mathbf{L}^2(\mathbf{T})} \le \|a\|_\infty \, \|p_n\|_{\mathbf{L}^2(\mathbf{T})} \quad \text{for all} \quad p_n \in \operatorname{im} P_n^{\mathbf{T}}, \tag{5.3}$$

where $\|a\|_\infty := \sup\{|a(t)| : t \in \mathbb{T}\}$. Finally, let us remark that

$$\lim_{n\to\infty} \left\| M_n^{\mathbf{T}} f - f \right\|_{\mathbf{L}^2(\mathbf{T})} = 0$$

for all bounded Riemann integrable functions $f : \mathbb{T} \to \mathbb{C}$.

Now let us consider the relations between some operators acting on functions defined on the interval and corresponding operators on the unit circle, which can be described by means of the transformation F. For a function $a : [-1, 1] \to \mathbb{C}$ we define the function $\widehat{a} : \mathbb{T} \to \mathbb{C}$ by

$$\widehat{a}(e^{is}) = a(\cos s).$$

Furthermore, by T we denote the orthogonal projection from $\mathbf{L}^2(\mathbb{T})$ onto $\mathbf{L}^2_{\mathrm{odd}}(\mathbb{T})$. Note that

$$T = \frac{1}{2}(I - W),$$

where

$$(Wf)(t) = f(t^{-1}), \quad t \in \mathbb{T}.$$

Lemma 5.3 (cf. [32], Lemma 5.2) *The following identities hold:*

(i) $aI = F^{-1}\widehat{a}F \qquad$ *for* $a \in \mathbf{L}^\infty[-1,1]$,

(ii) $\widetilde{L_n^\sigma} = F^{-1}M_n^{\mathbf{T}}F$,

(iii) $P_n^\sigma = F^{-1}P_n^{\mathbf{T}}FP_n^\sigma$,

(iv) $V^* = F^{-1}(e_{-1}P_{\mathbf{T}} + e_1Q_{\mathbf{T}})F$,

(v) $P_n^\sigma F^{-1}TP_n^{\mathbf{T}} = P_n^\sigma F^{-1}T$,

(vi) $P_n^{\mathbf{T}}FP_n^\sigma = FP_n^\sigma$,

(vii) $P_n^\sigma F^{-1}TW_n^{\mathbf{T}} = W_n^\sigma F^{-1}T(e_1P_{\mathbf{T}} + Q_{\mathbf{T}}e_1I)$,

(viii) $W_n^{\mathbf{T}}FP_n^\sigma = e_{-1}FW_n^\sigma$.

In the special case $\sigma = \varphi^{-1}$ *moreover the relation*

$$S = -F^{-1}\psi S_{\mathbf{T}}F$$

is valid.

Proof. If $a : [-1,1] \to \mathbb{C}$ and $u \in \mathbf{L}^2_\sigma$, then

$$(Fau)(e^{is}) = \sqrt{\pi}\psi(e^{is})a(\cos s)w_\sigma(\cos s)u(\cos s) = \widehat{a}(e^{is})(Fu)(e^{is}),$$

which implies the first relation.

Since for $u : [-1,1] \to \mathbb{C}$ we have $M_n^{\mathbf{T}}Fu \in \mathbf{L}^2_{\mathrm{odd}}(\mathbf{T}) \cap \operatorname{im} P_n^{\mathbf{T}}$ (see Lemma 5.2) and, for $j = 1, \ldots, n$,

$$(M_n^{\mathbf{T}}Fu)(t_{jn}^M) = (Fu)(t_{jn}^M) = \sqrt{\pi}w_\sigma(x_{jn}^\varphi)u(x_{jn}^\varphi),$$

it follows $F^{-1}M_n^{\mathbf{T}}Fu \in \operatorname{im} P_n^\sigma$ and $(F^{-1}M_n^{\mathbf{T}}Fu)(x_{jn}^\varphi) = u(x_{jn}^\varphi)$, $j = 1, \ldots, n$, and the second assertion is proved. The remaining assertions can be obtained by a straightforward computation if we remember the relation (see [20])

$$S_{\mathbf{T}}u = \sum_{n=0}^{\infty}\langle u, e_n\rangle e_n - \sum_{n=-\infty}^{-1}\langle u, e_n\rangle e_n \tag{5.4}$$

and its implications

$$P_{\mathbf{T}}u = \sum_{n=0}^{\infty}\langle u, e_n\rangle e_n, \qquad Q_{\mathbf{T}}u = \sum_{n=-\infty}^{-1}\langle u, e_n\rangle e_n.$$

To show the assertion concerning the special case $\sigma = \varphi^{-1}$, we note that relation (5.4) implies

$$(S_{\mathbf{T}}F\widetilde{u}_n)(e^{is}) = \frac{1}{\sqrt{2}i}\left(S_{\mathbf{T}}(e_{n+1} - e_{-n-1})\right)(e^{is})$$

$$= \frac{1}{\sqrt{2}i}\left(e^{i(n+1)s} + e^{-i(n+1)s}\right) = -\sqrt{2}i\cos(n+1)s\,.$$

On the other hand, we have

$$(FS\widetilde{u}_n)(e^{is}) = i(FT_{n+1})(e^{is}) = \sqrt{2}i\,\psi(e^{is})\cos(n+1)s\,,$$

which completes the proof. ■

In [31], a collocation method for the approximate solution of Cauchy singular integral equations

$$A^{\mathbf{T}}u := (aI + bS_{\mathbf{T}})u = f, \qquad f \in \mathbf{L}^2(\mathbb{T}) \text{ given, } u \in \mathbf{L}^2(\mathbb{T}) \text{ sought,}$$

on the unit circle is considered, where the operator $A^{\mathbf{T}} = aI + bS_{\mathbf{T}}$ is replaced by the approximate operators

$$A_n^{\mathbf{T}} = M_n^{\mathbf{T}}(aI + bS_{\mathbf{T}})P_n^{\mathbf{T}}$$

and an approximate solution is sought in $\operatorname{im} P_n^{\mathbf{T}}$. The investigation of this collocation method is based on the Banach algebra methods described in subsection 3.2, where $P_n^{\mathbf{T}}$ and $W_n^{\mathbf{T}}$ play the roles of P_n and W_n, respectively, and the corresponding sequence algebra $\mathcal{A}^{\mathbf{T}}$ and the ideal $\mathcal{I}^{\mathbf{T}} \subset \mathcal{A}^{\mathbf{T}}$ are defined by means of these operators.

Lemma 5.4 *Let $\{B_n^{\mathbf{T}}\} \in \mathcal{A}^{\mathbf{T}}$. Then the sequence $\{B_n\}$ defined by*

$$B_n = P_n^\sigma F^{-1} T B_n^{\mathbf{T}} F P_n^\sigma$$

belongs to the algebra $\mathcal{A}$.

Proof. Since $T^* = T$ and the adjoint of $F \in \mathcal{L}(\mathbf{L}^2_\sigma, \mathbf{L}^2_{\mathrm{odd}}(\mathbf{T}))$ is just F^{-1}, we have

$$B_n^* = P_n^\sigma F^{-1} T (B_n^{\mathbf{T}})^* T F P_n^\sigma,$$

hence B_n and B_n^* are evidently convergent. Now consider $\widetilde{B_n} = W_n^\sigma B_n W_n^\sigma$. We have

$$\widetilde{B_n} = W_n^\sigma F^{-1} T W_n^{\mathbf{T}}\; W_n^{\mathbf{T}} B_n^{\mathbf{T}} W_n^{\mathbf{T}}\; W_n^{\mathbf{T}} F W_n^\sigma,$$

and a short computation shows that

$$W_n^\sigma F^{-1} T W_n^{\mathbf{T}} = P_n^\sigma F^{-1} T e_1 I, \qquad W_n^{\mathbf{T}} F W_n^\sigma = e_{-1} F P_n^\sigma.$$

Evidently, the latter two expressions together with their adjoints are strongly convergent, which proves the assertion. ■

As a covering system of localizing classes the system $\{M_\tau^{\mathbf{T}}\}_{\tau\in\mathbf{T}}$ is chosen, where

$$M_\tau^{\mathbf{T}} = \left\{\{M_n^{\mathbf{T}} f P_n^{\mathbf{T}}\} + \mathcal{I}^{\mathbf{T}} : f \in m_\tau^{\mathbf{T}}\right\},$$

and $m_\tau^{\mathbf{T}}$ is defined analogously to m_τ.

The coefficients a and b of the operator $A^{\mathbf{T}}$ are assumed to be piecewise continuous on $\mathbb{T}$. We point out again that the values of piecewise continuous functions are uniquely determined by the behaviour of the function in an (arbitrarily small) neighbourhood of the argument (cf. Section 1).

Lemma 5.5 *Let a, b be Riemann integrable, even functions on $\mathbb{T}$, that is,*

$$a(t) = a(\bar{t}), \qquad b(t) = b(\bar{t}), \qquad t \in \mathbb{T}.$$

Then the operators $A_n^{\mathbf{T}} := M_n^{\mathbf{T}}(aI - \psi b S_{\mathbf{T}})P_n^{\mathbf{T}}$ commute with $TFP_n^{\sigma}F^{-1}T$.

Proof. Relation (5.4) and Lemma 5.2 show that the space $\operatorname{im} P_n^{\mathbf{T}} \cap \mathbf{L}^2_{\text{odd}}(\mathbb{T})$ remains invariant under the action of $A_n^{\mathbf{T}}$. The operator $TFP_n^{\sigma}F^{-1}T$ is just the orthogonal projection from $\mathbf{L}^2(\mathbb{T})$ onto this space. Hence, we have

$$TFP_n^{\sigma}F^{-1}T\, A_n^{\mathbf{T}}\, TFP_n^{\sigma}F^{-1}T = A_n^{\mathbf{T}}\, TFP_n^{\sigma}F^{-1}T.$$

We pass over to the adjoint operator (note the remarks in the proof of Lemma 5.4) and get

$$TFP_n^{\sigma}F^{-1}T\, (A_n^{\mathbf{T}})^*\, TFP_n^{\sigma}F^{-1}T = TFP_n^{\sigma}F^{-1}T\, (A_n^{\mathbf{T}})^*. \tag{5.5}$$

In [31] the relation

$$(A_n^{\mathbf{T}})^* = M_n^{\mathbf{T}}\bar{a}P_n^{\mathbf{T}} - S_{\mathbf{T}}M_n^{\mathbf{T}}\psi\bar{b}P_n^{\mathbf{T}}$$

is shown, and evidently this operator also maps the space $\operatorname{im} P_n^{\mathbf{T}} \cap \mathbf{L}^2_{\text{odd}}(\mathbb{T})$ into itself, that is,

$$TFP_n^{\sigma}F^{-1}T\, (A_n^{\mathbf{T}})^*\, TFP_n^{\sigma}F^{-1}T = (A_n^{\mathbf{T}})^*\, TFP_n^{\sigma}F^{-1}T.$$

If we compare this relation with (5.5) and again pass over to the adjoint operators, we obtain the assertion. ■

The following proposition summarizes some of the main results of [31].

Proposition 5.6 ([31]) *Assume that a and b are bounded Riemann integrable functions.*

(a) Then the sequence $\{M_n^{\mathbf{T}}(aI + bS_{\mathbf{T}})P_n^{\mathbf{T}}\}$ belongs to the algebra $\mathcal{A}^{\mathbf{T}}$.

(b) For every continuous function $f : \mathbb{T} \longrightarrow \mathbb{C}$ there exist compact operators $K_1, K_2 \in \mathcal{K}(\mathbf{L}^2(\mathbb{T}))$ such that

$$M_n^{\mathbf{T}} f P_n^{\mathbf{T}} M_n^{\mathbf{T}}(aI + bS_{\mathbf{T}})P_n^{\mathbf{T}} - M_n^{\mathbf{T}}(aI + bS_{\mathbf{T}})P_n^{\mathbf{T}} M_n^{\mathbf{T}} f P_n^{\mathbf{T}}$$

$$= P_n^{\mathbf{T}} K_1 P_n^{\mathbf{T}} + W_n^{\mathbf{T}} K_2 W_n^{\mathbf{T}} + C_n$$

with $\lim_{n\to\infty} \|C_n\|_{\mathcal{L}(\mathbf{L}^2(\mathbf{T}))} = 0$. In particular, the cosets $\{M_n^{\mathbf{T}}(aI + bS_{\mathbf{T}})P_n^{\mathbf{T}}\} + \mathcal{I}^{\mathbf{T}}$ commute with all elements of $\bigcup_{\tau\in\mathbf{T}} M_\tau^{\mathbf{T}}$.

(c) For $a, b \in \mathbf{PC}(\mathbb{T})$ the sequence $\{M_n^{\mathbf{T}}(aI + bS_{\mathbf{T}})P_n^{\mathbf{T}}\}$ is stable if and only if the operator $aI + bS_{\mathbf{T}}$ is invertible in $\mathcal{L}(\mathbf{L}^2(\mathbb{T}))$.

We conclude this subsection with a proposition that states a connection between the operator sequences occurring in the collocation methods on the interval and on the unit circle, respectively.

Proposition 5.7 *Let $a \in \mathbf{PC}[-1,1]$, $b \in \mathbf{PC}_0[-1,1]$ and assume that we do not have one of the special cases $\alpha = \frac{1}{2}$ or $\beta = \frac{1}{2}$. Then we have*

$$\{\widetilde{L_n^\sigma}(aI + bS)P_n^\sigma\} = \left\{F^{-1}\left[M_n^{\mathbf{T}}(\widehat{a}I - \psi\widehat{b}S_{\mathbf{T}})P_n^{\mathbf{T}}\right]FP_n^\sigma\right\} + \{C_n\},$$

where $\{C_n\} \in \mathcal{I}$. In the special case $\sigma = \varphi^{-1}$, we have moreover the relation

$$\{\widetilde{L_n^\sigma}(aI + bS)P_n^\sigma\} = \left\{F^{-1}\left[M_n^{\mathbf{T}}(\widehat{a}I - \psi\widehat{b}S_{\mathbf{T}})P_n^{\mathbf{T}}\right]FP_n^\sigma\right\}$$

for arbitrary $a\,, b \in \mathbf{PC}[-1,1]$.

Proof. The relation $\widetilde{L_n^\sigma} aP_n^\sigma = F^{-1}M_n^{\mathbf{T}}\widehat{a}P_n^{\mathbf{T}}FP_n^\sigma$ is an immediate consequence of Lemma 5.3. Hence, we can confine ourselves to the consideration of $\widetilde{L_n^\sigma} bSP_n^\sigma$.

First assume that $b = \chi b_1$, where χ is a Riemann integrable function (for instance the characteristic function of some subinterval) and $b_1 \in \mathbf{C}_{0,0}^{1,\eta}$ with a real number $\eta > \max\left\{\frac{1}{4} + \frac{\alpha}{2}, \frac{1}{4} + \frac{\beta}{2}, 0\right\}$ (Compare the proof of Proposition 4.22). Equation (4.5) now shows that

$$\chi b_1 S = K + i\chi b_1\varphi^{-1}(eI - V^*)$$

with $K \in \mathcal{K}(\mathbf{L}_\sigma^2, \mathbf{R})$ (and therefore $K \in \mathcal{K}(\mathbf{L}_\sigma^2)$), where $\mathbf{R}$ denotes the space of Riemann integrable functions (cf. Lemma 4.13). Hence, in view of Lemma 3.10 and Corollary 4.5 we get

$$\left\|\widetilde{L_n^\sigma} KP_n^\sigma - K\right\|_{\mathcal{L}(\mathbf{L}_\sigma^2)} \to 0 \qquad (n \to \infty).$$

As a consequence, we have $\widetilde{L_n^\sigma} KP_n^\sigma = P_n^\sigma KP_n^\sigma + C_n$, where $\|C_n\|_{\mathcal{L}(\mathbf{L}_\sigma^2)} \to 0$, that is, $\{\widetilde{L_n^\sigma} KP_n^\sigma\} \in \mathcal{I}$.

Now let us consider the remaining term. Lemma 5.3 yields the relation

$$\begin{aligned}
&\widetilde{L_n^\sigma}\left(i\chi b_1\varphi^{-1}(eI - V^*)\right)P_n^\sigma \\
&= F^{-1}\left[M_n^{\mathbf{T}}\left(i\widehat{\chi}\widehat{b_1}\widehat{\varphi}^{-1}(\widehat{e}I - (e_{-1}P_{\mathbf{T}} + e_1Q_{\mathbf{T}}))\right)P_n^{\mathbf{T}}\right]FP_n^\sigma \\
&= F^{-1}\left[M_n^{\mathbf{T}}(cP_{\mathbf{T}} + \widetilde{c}Q_{\mathbf{T}})P_n^{\mathbf{T}}\right]FP_n^\sigma
\end{aligned}$$

with $c = i\widehat{\chi}\widehat{b_1}\widehat{\varphi}^{-1}(\widehat{e} - e_{-1})$, that is,

$$c(x+iy) = \begin{cases} -(\chi b_1)(x), & y \geq 0 \\ (\chi b_1)(x), & y < 0, \end{cases}$$

and $\widetilde{c}(s) = c(s^{-1})$. Thus, the first assertion is proved for coefficients of the form $b = \chi b_1$ with $\chi \in \mathbf{R}$ and $b \in \mathbf{C}_{0,0}^{1,\eta}$. By approximation, one can pass over first to piecewise continuous coefficients with a finite number of jumps and then to arbitrary $b \in \mathbf{PC}_0$ (Remember Proposition 4.6, relation (5.3), and the fact that $\mathcal{I}$ is a closed subset of $\mathcal{A}$, and compare the proof of Proposition 4.22).

The assertion concerning the special case $\sigma = \varphi^{-1}$ can be easily derived from Lemma 5.3. ■

Remark 5.8 *We point out that for the validity of the transformation it is essential to define* ψ *just the way we did in (5.1). Hence, if not* $b(\pm 1) = 0$, ψ *is not an element of* $\mathbf{PC}(\mathbb{T})$ *in the sense of the definition given in Section 1, and Lemma 5.6(c) does not apply. (The crucial point is the fact that the points* ± 1, *in which* ψ *violates the definition of a* **PC**-*function, belong to the nodes of the interpolation operator* $M_n^{\mathbf{T}}$, *and thus the values* $\psi(\pm 1)$ *cause a global change of the resulting interpolation polynomial.) Consider for instance* $A = S \in \mathcal{L}(\mathbf{L}_\sigma^2)$, *which is not invertible. Then* $\{A_n\} = \{-F^{-1}M_n^{\mathbf{T}}\psi S_{\mathbf{T}}P_n^{\mathbf{T}}FP_n\}$ *is not stable. If we, however, modified the function* ψ *in* ± 1 *to obtain the function* $\varrho \in \mathbf{PC}(\mathbb{T})$,

$$\varrho(e^{is}) = \begin{cases} 1 \ , & 0 < s \le \pi \, , \\ -1 \ , & \pi < s \le 2\pi \, , \end{cases}$$

we would arrive at $\{M_n^{\mathbf{T}}\varrho S_{\mathbf{T}}P_n^{\mathbf{T}}\}$ *which is stable due to Lemma 5.6(c). (The uniformly bounded inverses are* $\{S_{\mathbf{T}}M_n^{\mathbf{T}}\varrho P_n^{\mathbf{T}}\}$.) *This shows that one cannot simply reduce the stability of* $\{A_n\}$ *to the stability of a corresponding collocation method on the unit circle. (The latter statement also applies to the case* $b(\pm 1) = 0$, *since the transformation in Prop. 5.7 is, in general, only valid modulo the ideal* $\mathcal{I}$.) *The transformation to the unit circle can merely be employed as an auxiliary tool in the approach based on Theorem 3.12 and the local principle.*

Remark 5.9 *If we have one of the special cases with respect to the exponents of the weight* σ *considered in Subsection 4.3, we slightly modify the assumptions of the preceding lemma according to Remark 4.23, which allows the approximation of* $\{\widetilde{L_n^\sigma bSP_n^\sigma}\}$ *in* $\mathcal{A}$, *where* $b \in \mathbf{PC}_0$ *with a finite number of jumps, by corresponding sequences with piecewise smooth coefficients.*

5.2 Main result

Now we are in a position to derive the main stability result by considering the invertibility of the coset $\{A_n\} + \mathcal{I}$. As a preliminary result we need the commutativity of this coset with all elements of the localizing classes.

Lemma 5.10 (cf. [33], Lemma 6.1) *Let* $a \in \mathbf{PC}[-1,1]$, $b \in \mathbf{PC}_0[-1,1]$ *(in the special case* $\sigma = \varphi^{-1}$ *it is sufficient that* $b \in \mathbf{PC}$). *Then the coset* $\{\widetilde{L_n^\sigma(aI+bS)P_n^\sigma}\} + \mathcal{I}$ *commutes with all elements of* $\bigcup_{\tau \in [-1,1]} M_\tau$.

Proof. Let $\tau \in [-1,1]$ and $f \in m_\tau$. In view of Propositions 5.7 and 5.6(b) there exist compact operators $K_1, K_2 \in \mathcal{K}(\mathbf{L}^2(\mathbb{T}))$ and a sequence $\{C_n\} \in \mathcal{A}^{\mathbf{T}}$ with $\lim_{n\to\infty} \|C_n\|_{\mathcal{L}(\mathbf{L}^2(\mathbf{T}))} = 0$ such that

$$\{\widetilde{L_n^\sigma} f P_n^\sigma\}\{\widetilde{L_n^\sigma}(aI+bS)P_n^\sigma\} - \{\widetilde{L_n^\sigma}(aI+bS)P_n^\sigma\}\{\widetilde{L_n^\sigma} f P_n^\sigma\} + \mathcal{I}$$

$$= \{F^{-1}\left[M_n^{\mathbf{T}}\widehat{f}P_n^{\mathbf{T}}M_n^{\mathbf{T}}(\widehat{a}I - \widehat{b}\psi S_{\mathbf{T}})P_n^{\mathbf{T}} - M_n^{\mathbf{T}}(\widehat{a}I - \widehat{b}\psi S_{\mathbf{T}})P_n^{\mathbf{T}}M_n^{\mathbf{T}}\widehat{f}P_n^{\mathbf{T}}\right]FP_n^\sigma\} + \mathcal{I}$$

$$= \{P_n^\sigma F^{-1} T(P_n^{\mathbf{T}} K_1 P_n^{\mathbf{T}} + W_n^{\mathbf{T}} K_2 W_n^{\mathbf{T}} + C_n) F P_n^\sigma\} + \mathcal{I}.$$

(We insert the projection T to be able to consider the three summands individually, since it is not guaranteed that each summand inside the parentheses maps $\operatorname{im} F$ into $\operatorname{im} F$.) If we use the relations given in Lemma 5.3, we see that the latter expression in braces equals

$$\{P_n^\sigma F^{-1} T K_1 F P_n^\sigma + W_n^\sigma F^{-1} T(e_1 P_{\mathbf{T}} + Q_{\mathbf{T}} e_1 I) K_2 e_{-1} F W_n^\sigma + P_n^\sigma F^{-1} T C_n F P_n^\sigma\},$$

which is obviously an element of $\mathcal{I}$. ■

Now we are able to give local representatives for $\{A_n\} + \mathcal{I}$.

Lemma 5.11 *Let $\tau \in [-1,1]$, $a, a_\tau \in \mathbf{PC}$, $b, b_\tau \in \mathbf{PC}_0$ such that*

$$a_\tau(\tau \pm 0) = a(\tau \pm 0), \qquad b_\tau(\tau \pm 0) = b(\tau \pm 0). \tag{5.6}$$

Then we have

$$\{\widetilde{L_n^\sigma}(aI+bS)P_n^\sigma\} + \mathcal{I} \overset{M_\tau}{\sim} \{\widetilde{L_n^\sigma}(a_\tau I + b_\tau S)P_n^\sigma\} + \mathcal{I}.$$

In the case $\sigma = \varphi^{-1}$ we only require $b, b_\tau \in \mathbf{PC}$.

Proof. Let $f \in m_\tau$. We have

$$\left\|\{\widetilde{L_n^\sigma} f P_n^\sigma\}\{\widetilde{L_n^\sigma}((a-a_\tau)I + (b-b_\tau)S)P_n^\sigma\}\right\|_{\mathcal{A}/\mathcal{I}}$$

$$\le \|\widetilde{L_n^\sigma} f (a-a_\tau) P_n^\sigma + \widetilde{L_n^\sigma} f(b-b_\tau)P_n^\sigma\, \widetilde{L_n^\sigma} S P_n^\sigma\|_{\mathcal{L}(\mathbf{L}^2_\sigma)}$$

$$\le \|f(a-a_\tau)\|_\infty + \text{const}\ \|f(b-b_\tau)\|_\infty,$$

which can be made arbitrarily small by a suitable choice of f. Thus, we have proved the assertion (note Lemma 5.10). ■

Remark 5.12 *In the special cases with respect to the exponents of the weight σ considered in Subsection 4.3, we slightly modify the preceding lemma and its proof. If for instance $\alpha = \frac{1}{2}$, we require that $b(x)$ and $b_\tau(x)$ are of the order $o((1-x)^\eta)$ for $x \to 1$, where $\eta > \frac{1+\max\{\alpha,\beta,0\}}{2}$, which ensures $\|\widetilde{L_n^\sigma}(1-x)^\eta S P_n^\sigma\| \le const$. Then we estimate*

$$\|\widetilde{L_n^\sigma} f(b-b_\tau)P_n^\sigma\, \widetilde{L_n^\sigma} S P_n^\sigma\| \le \|f(1-x)^{-\eta}(b-b_\tau)\|_\infty\, \|\widetilde{L_n^\sigma}(1-x)^\eta S P_n^\sigma\|.$$

At this point we need some material on the invertibility of singular integral operators.

Definition 5.13 *Let $a, b \in \mathbf{PC}[-1,1]$ and assume that*

$$\inf_{x\in[-1,1]} |a(x)+b(x)| > 0 \qquad \text{and} \qquad \inf_{x\in[-1,1]} |a(x)-b(x)| > 0. \tag{5.7}$$

To the operator $aI + bS \in \mathcal{L}(\mathbf{L}^2_\sigma)$, $\sigma = v^{\alpha,\beta}$, we assign the symbol $\mathbf{c} : [-1,1]\times[0,1] \to \mathbb{C}$ defined by

$$\mathbf{c}(x,\mu) := \begin{cases} c(x-0)\mu + c(x+0)(1-\mu), & \mu\in[0,1]\,,\ x\in(-1,1)\,, \\ c(1)[1-f_\alpha(\mu)] + f_\alpha(\mu), & \mu\in[0,1]\,,\ x=1\,, \\ 1-f_\beta(\mu)+c(-1)f_\beta(\mu), & \mu\in[0,1]\,,\ x=-1\,, \end{cases} \tag{5.8}$$

where $c(x) = \dfrac{a(x)+b(x)}{a(x)-b(x)}$ and

$$f_\alpha(\mu) = \begin{cases} \frac{\sin\alpha\pi\mu}{\sin\alpha\pi}\, e^{-i\alpha\pi(\mu-1)}, & \alpha\neq 0, \\ \mu, & \alpha = 0, \end{cases}$$

and f_β is defined analogously. Note that in the special case $\alpha = -\frac{1}{2}$ the expression $z_1[1-f_\alpha(\mu)] + z_2 f(\mu)$, $\mu\in[0,1]$, describes the half-circle line from z_1 to z_2 that lies to the right of the straight line from z_1 to z_2.
If $a, b \in \mathbf{PC}(\mathbb{T})$ and

$$\inf_{t\in\mathbf{T}} |a(t)+b(t)| > 0 \qquad \text{and} \qquad \inf_{t\in\mathbf{T}} |a(t)-b(t)| > 0, \tag{5.9}$$

we define a symbol $\mathbf{c}^{\mathbf{T}} : \mathbb{T}\times[0,1] \to \mathbb{C}$ for $aI + bS_{\mathbf{T}} \in \mathcal{L}(\mathbf{L}^2(\mathbb{T}))$ by

$$\mathbf{c}^{\mathbf{T}}(t,\mu) := c(t-0)\mu + c(t+0)(1-\mu), \qquad \mu\in[0,1]\,,\ t\in\mathbb{T},$$

where $c(t) = \frac{a(t)+b(t)}{a(t)-b(t)}$. Thus, the images of $\mathbf{c}$ and $\mathbf{c}^{\mathbf{T}}$ are closed curves in the complex plane that possess a natural orientation, and by wind $\mathbf{c}(x,\mu)$ *and* wind $\mathbf{c}^{\mathbf{T}}(t,\mu)$ *we denote the winding number of these curves with respect to the origin 0.*

Proposition 5.14 ([20], Theorems IX.3.1, IX.4.1) *Let $a, b \in \mathbf{PC}[-1,1]$. Then the operator $aI + bS$ is Fredholm in $\mathbf{L}^2_\sigma$ if and only if (5.7) is valid and $\mathbf{c}(x,\mu) \neq 0$ for all $x\in[-1,1]$, $\mu\in[0,1]$. In this case the operator is invertible only from the left, invertible only from the right or invertible from both sides if* wind $\mathbf{c}$ *is positive, negative, or zero, respectively, and its index equals $\kappa = -$ wind $\mathbf{c}$. If $a, b \in \mathbf{PC}(\mathbb{T})$, an analogous relation holds for the invertibility of $aI + bS_{\mathbf{T}}$ in $\mathbf{L}^2(\mathbb{T})$ with respect to condition (5.9) and the behaviour of $\mathbf{c}^{\mathbf{T}}$.*

Now we are able to prove the main result. In the case $b(\pm 1) = 0$ this can be done on the basis of Proposition 5.7 without using the local principle.

Theorem 5.15 *Let $a \in \mathbf{PC}[-1,1]$, $b \in \mathbf{PC}_0[-1,1]$. Then the sequence $\{A_n\} = \{\widetilde{L_n^\sigma}(aI+bS)P_n^\sigma\}$ is stable if and only if the operator $A = aI + bS$ is invertible in $\mathbf{L}^2_\sigma$.*

Proof. Due to Theorem 3.12 and Remark 4.24 it only remains to consider the invertibility of the coset $\{A_n\} + \mathcal{I}$. The invertibility of $aI + bS$ (and hence also of $aI - bS$) in $\mathbf{L}^2_\sigma$ implies the invertibility of the operator $A^{\mathbf{T}} := \widehat{a}I - \widehat{\psi b}S_{\mathbf{T}}$ in $\mathbf{L}^2(\mathbb{T})$. (Note that the coefficients of this operator are in $\mathbf{PC}(\mathbb{T})$ because of $b(\pm 1) = 0$.) Lemma 5.6(c) now shows that the sequence $\{M_n^{\mathbf{T}} A^{\mathbf{T}} P_n^{\mathbf{T}}\}$ is stable in $\mathbf{L}^2(\mathbb{T})$, which means that there exists a sequence $B_n^{\mathbf{T}} \in \mathcal{L}(\operatorname{im} P_n^{\mathbf{T}})$ such that

$$A_n^{\mathbf{T}} B_n^{\mathbf{T}} = B_n^{\mathbf{T}} A_n^{\mathbf{T}} = P_n^{\mathbf{T}}$$

for all sufficiently large n and $\|B_n^{\mathbf{T}} P_n^{\mathbf{T}}\|_{\mathcal{L}(\mathbf{L}^2(\mathbf{T}))} \le \text{const}$. By virtue of Proposition 5.7 we have

$$\{A_n\} + \mathcal{I} = \{F^{-1} A_n^{\mathbf{T}} F P_n^\sigma\} + \mathcal{I},$$

where $A_n^{\mathbf{T}} = M_n^{\mathbf{T}} A^{\mathbf{T}} P_n^{\mathbf{T}}$. Now we put

$$B_n := P_n^\sigma F^{-1} T B_n^{\mathbf{T}} F P_n^\sigma .$$

(We insert the projection T since it is not clear whether the image of $B_n^{\mathbf{T}}$ is contained in $\mathbf{L}^2_{\text{odd}}(\mathbb{T})$. For the same reason it is necessary to multiply by P_n^σ from the left in order to make sure that $B_n \in \mathcal{L}(\operatorname{im} P_n^\sigma)$.) Due to Lemma 5.4 we have $\{B_n\} \in \mathcal{A}$. Using the relations from Lemma 5.3 we obtain

$$B_n A_n = F^{-1} T B_n^{\mathbf{T}} F P_n^\sigma F^{-1} A_n^{\mathbf{T}} P_n^{\mathbf{T}} F P_n^\sigma + C_n = P_n^\sigma + C_n,$$

where $\{C_n\} \in \mathcal{I}$, hence $\{A_n\} + \mathcal{I}$ is invertible from the right. Further, we have in view of Lemma 5.5

$$\begin{aligned} A_n B_n &= F^{-1} A_n^{\mathbf{T}}\, T F P_n^\sigma F^{-1} T\, B_n^{\mathbf{T}} F P_n^\sigma + D_n \\ &= F^{-1}\, T F P_n^\sigma F^{-1} T\, A_n^{\mathbf{T}} B_n^{\mathbf{T}} F P_n^\sigma + D_n = P_n^\sigma + D_n, \end{aligned}$$

where $\{D_n\} \in \mathcal{I}$. Hence, $\{A_n\} + \mathcal{I}$ is invertible, and the proof is complete. ∎

In the special case $\sigma = \varphi^{-1}$ we can dispense with the condition $b(\pm 1) = 0$, but we can not make use of Proposition 5.6(c) immediately, since the coefficient $\widehat{\psi b}$ is not in $\mathbf{PC}(\mathbb{T})$ if b does not vanish in ± 1. At this point, the application of the local principle becomes inevitable. The following lemma will enable us to cope with certain local representatives in ± 1 via a Neumann series argument.

Lemma 5.16 ([33], Lemma 6.3) *Let $\sigma = \varphi^{-1}$. Assume that $a, b \in \mathbf{PC}[-1,1]$ and that $A = aI + bS$ and $\widetilde{A} = aI - bS$ are invertible in $\mathbf{L}^2_\sigma$. Then $|a(-1)| > |b(-1)|$ and $|a(1)| > |b(1)|$.*

Proof. The invertibility of A implies $\operatorname{wind} \mathbf{c}(x,\mu) = 0$ in view of Proposition 5.14. Since $\widetilde{A}$ is also assumed to be invertible in $\mathbf{L}^2_\sigma$, analogous relations hold for $\mathbf{d}(x,\mu)$, which is defined by (5.8) with $d(x) := 1/c(x)$ instead of $c(x)$. We will show now that under the assumptions of the lemma both $c(1)$ and $c(-1)$ are located in the right half plane, from which the assertion of the lemma follows immediately. Evidently, the real parts of $c(1)$ and $c(-1)$ cannot vanish because of $\mathbf{c}(\pm 1,\mu) \neq 0$ and $\mathbf{d}(\pm 1,\mu) \neq 0$, $\mu \in [0,1]$. Consider for instance the case $\operatorname{Re} c(1) < 0$, $\operatorname{Re} c(-1) > 0$. By arg we denote a continuous branch of the argument defined on $\{\mathbf{c}(x,\mu) : -1 < x < 1,\ \mu \in [0,1]\} \cup \{c(-1), c(1)\}$. Let $c(1) = |c(1)| \exp(i\zeta)$, $c(-1) = |c(-1)| \exp(i\eta)$, $\frac{\pi}{2} < \zeta < \frac{3\pi}{2}$, $-\frac{\pi}{2} < \eta < \frac{\pi}{2}$. Then the argument increase of the closed curve described by $\mathbf{c}(x,\mu)$ equals $2\pi - \zeta + \eta + \arg c(1) - \arg c(-1)$, which must be zero. On the other hand, the argument increase of $\mathbf{d}(x,\mu)$ is $\zeta - \eta - \arg c(1) + \arg c(-1) = 2\pi$ in contradiction to the invertibility of $\widetilde{A}$. All other cases can be treated analogously. ■

Theorem 5.17 ([33], Th. 6.4) *Let $\sigma = \varphi^{-1}$. Then for $a, b \in \mathbf{PC}[-1,1]$ the sequence $\{\widetilde{L^\sigma_n}(aI+bS)P^\sigma_n\}$ is stable in $\mathbf{L}^2_\sigma$ if and only if the operators $A = aI+bS$ and $\widetilde{A} = aI-bS$ are invertible in $\mathbf{L}^2_\sigma$.*

Proof. Again we investigate the invertibility of the coset $\{A_n\} + \mathcal{I}$. Here we do this on the basis of the local principle Theorem 3.4 by considering the invertibility of local representatives $\{A^\tau_n\} + \mathcal{I}$, $\tau \in [-1,1]$, which are chosen according to Lemma 5.11.

First let $\tau \in (-1,1)$. We choose $a_\tau \in \mathbf{PC}$, $b_\tau \in \mathbf{PC}_0$ such that equation (5.6) is satisfied and $a_\tau I + b_\tau S$ is invertible in $\mathbf{L}^2_\sigma$. Then we can show in the same way as in the proof of Theorem 5.15 that the coset $\{A^\tau_n\} + \mathcal{I}$, where $A^\tau_n = \widetilde{L^\sigma_n}(a_\tau I + b_\tau S)P^\sigma_n$, is invertible (and therefore of course also M_τ-invertible).

For $\tau = \pm 1$, we choose $A^\tau_n = \widetilde{L^\sigma_n}[a(\tau)I+b(\tau)S]P^\sigma_n$, which is contained in $\mathcal{A}$ (compare Subsection 4.6). In this case, the stability of the sequence

$$\{A^{\mathbf{T}}_{n,\tau}\} := \{M^{\mathbf{T}}_n[a(\tau)I - b(\tau)\psi S_{\mathbf{T}}]P^{\mathbf{T}}_n\} = \{a(\tau)[P^{\mathbf{T}}_n - M^{\mathbf{T}}_n(a(\tau))^{-1}b(\tau)\psi S_{\mathbf{T}}P^{\mathbf{T}}_n]\}$$

is easily seen by Lemma 5.16 and relation (5.3), and we can again continue as in the proof of Theorem 5.15. ■

6 The collocation method for systems of singular integral equations

6.1 Preliminaries

One of the advantages of our collocation method based on weighted polynomials is the fact that it can be easily generalized to the system case, that is, to problems of the form

$$\sum_{j=1}^{k}(a_{ij}I + b_{ij}S)u_j = f_i, \qquad i = 1, \dots, k, \tag{6.1}$$

where a_{ij}, $b_{ij} \in \mathbf{PC}$ and $f_i \in \mathbf{L}^2_\sigma$ are given functions and $u_j \in \mathbf{L}^2_\sigma$, $(j = 1, \dots, k)$ are unknown. The usual polynomial approximation methods based on the invariance relation in Proposition 1.1 are not suitable for this kind of problems if the matrices (a_{ij}), (b_{ij}) are not diagonal.

In the following considerations we will assume $v = \varphi$, that means we use weighted Chebyshev polynomials of the second kind, and for the sake of simplicity we will exclude the special cases $\alpha = \frac{1}{2}$ and $\beta = \frac{1}{2}$. Furthermore, we will restrict ourselves to the case $b_{ij}(\pm 1) = 0$, $i, j = 1, \dots, k$.

Let $k > 1$ be an integer. We put $X = (\mathbf{L}^2_\sigma)^k$, which denotes the cross product of k copies of the space $\mathbf{L}^2_\sigma$, equipped with the inner product

$$\langle \underline{u}\, , \underline{v}\, \rangle_{\sigma,k} := \sum_{j=1}^{k} \langle u_j, v_j \rangle_\sigma,$$

where $\underline{u} = (u_j)_{j=1}^k$, $\underline{v} = (v_j)_{j=1}^k \in (\mathbf{L}^2_\sigma)^k$. By $\mathbf{C}^{k\times k}$ we denote the set of all $k \times k$-matrices with entries from $\mathbf{C}[-1,1]$, endowed with the norm

$$\|\underline{a}\,\|_\infty := \left(k \sum_{i=1}^{k} \sum_{j=1}^{k} \|a_{ij}\|^2_\infty \right)^{\frac{1}{2}},$$

where $\underline{a} = (a_{ij})_{i,j=1}^k \in \mathbf{C}^{k\times k}$. For other spaces of scalar functions, the corresponding spaces of matrix-valued functions are defined in an analogous way. The norm introduced above is compatible with the norm in $X = (\mathbf{L}^2_\sigma)^k$: If $a \in \mathbf{C}^{k\times k}$, $u \in (\mathbf{L}^2_\sigma)^k$, we have

$$\|\underline{a}\,\underline{u}\,\|^2_X = \sum_{i=1}^{k} \left\| \sum_{j=1}^{k} a_{ij}u_j \right\|^2_\sigma \le k \sum_{i=1}^{k} \sum_{j=1}^{k} \|a_{ij}u_j\|^2_\sigma \le \|\underline{a}\,\|^2_\infty \|\underline{u}\,\|^2_X.$$

Hence, for the multiplication operator $\underline{a}\, I$ we have $\|\underline{a}\, I\|_{\mathcal{L}(X)} \le \|\underline{a}\,\|_\infty$. By $\underline{S}$ we denote the diagonal operator $(\delta_{ij}S)_{i,j=1}^k \in \mathcal{L}(X)$. Analogously, $\underline{P}^\sigma_n$, $\widetilde{\underline{L}^\sigma_n}$, $\underline{W}^\sigma_n$ etc. are defined. We have $X_n := \operatorname{im} \underline{P}^\sigma_n = \operatorname{span}\{\widetilde{\underline{u}}_{jm} : j = 0, \dots, n-1;\, m = 1, \dots, k\}$, where $\widetilde{\underline{u}}_{jm} = (\delta_{im}\widetilde{u}_j)_{i=1}^k$. Thus, we can write (6.1) in the form

$$(\underline{a}\, I + \underline{b}\, \underline{S}\,)\underline{u} = \underline{f}$$

($\underline{a}\,,\underline{b} \in \mathbf{C}^{k\times k}$ given, $\underline{u} \in X$ unknown), and we will consider the collocation method

$$\widetilde{\underline{L}_n^\sigma}(\underline{a}\,I + \underline{b}\,\underline{S}\,)\underline{P}_n^\sigma \underline{u}_n = \widetilde{\underline{L}_n^\sigma}\underline{f}$$

($\underline{u}_n \in X_n$) for its approximate solution.

6.2 Strong convergence

Now we again consider the strong convergence of the sequences $A_n = \widetilde{\underline{L}_n^\sigma}(\underline{a}\,I + \underline{b}\,\underline{S}\,)\underline{P}_n^\sigma$, A_n^*, $\widetilde{A_n}$ and $\widetilde{A_n}^*$. First we note that $\underline{P}_n^\sigma$ converges strongly to I on X and $\widetilde{\underline{L}_n^\sigma}$ converges to I on vector-valued functions that satisfy the conditions of Corollary 4.5 in every component.

Lemma 6.1 *We have* $\widetilde{\underline{L}_n^\sigma}\underline{a}\,\underline{P}_n^\sigma \to \underline{a}\,I$ *and* $\left\|\widetilde{\underline{L}_n^\sigma}\underline{a}\,\underline{P}_n^\sigma\right\| \le \|\underline{a}\,\|_\infty$ *if* $\underline{a} \in \mathbf{R}^{k\times k}$.

Proof. To show the convergence on a dense subset, consider $\widetilde{\underline{L}_n^\sigma}\underline{a}\,\widetilde{\underline{u}}_{jm} = \left(\widetilde{L_n^\sigma}a_{im}\widetilde{u}_j\right)_{i=1}^k$, which converges componentwise and therefore in $(\mathbf{L}_\sigma^2)^k$. For the uniform boundedness, we estimate

$$\left\|\widetilde{\underline{L}_n^\sigma}\underline{a}\,\underline{P}_n^\sigma\underline{u}\right\|_X^2 = \sum_{i=1}^k \left\|\widetilde{L_n^\sigma}\sum_{j=1}^k a_{ij}P_n^\sigma u_j\right\|_\sigma^2 \le k\sum_{i=1}^k\sum_{j=1}^k \left\|\widetilde{L_n^\sigma}a_{ij}P_n^\sigma u_j\right\|_\sigma^2$$

$$\le k\sum_{i=1}^k\sum_{j=1}^k \|a_{ij}\|_\infty^2 \left\|u_j^2\right\|_\sigma \le \|\underline{a}\,\|_\infty^2\|\underline{u}\,\|_X^2. \quad \blacksquare$$

Proposition 4.8 can be taken over without any changes with respect to the dense subset $\operatorname{span}\{\widetilde{\underline{u}}_{jm} : j = 0, 1, \ldots;\ m = 1, \ldots, k\}$.

Moreover, a decomposition analogous to (4.1) is possible with *scalar-valued* functions $\widetilde{a}$ and w. Therefore, we have $K = w^{-1}(w\underline{S} - \underline{S}\,wI) \in \underline{\mathcal{K}}((\mathbf{L}_\sigma^2)^k, (\mathbf{C}^{0,\lambda})^k)$ for some $\lambda > 0$, and hence $\|\widetilde{\underline{L}_n^\sigma}K\underline{P}_n^\sigma - K\|_{\mathcal{L}(X)} \to 0$, in particular $\|\widetilde{\underline{L}_n^\sigma}K\underline{P}_n^\sigma\|_{\mathcal{L}(X)} \le const.$

If we finally note that we can show in the same way as before the uniform boundedness of $\widetilde{\underline{L}_n^\sigma}w^{-1}(\widetilde{a}I + i\underline{S}\,)w\underline{P}_n^\sigma$, we arrive at the following result:

Lemma 6.2 *If* $\underline{a}\,,\,\underline{b} \in \mathbf{R}^{k\times k}$, *we have* $A_n = \widetilde{\underline{L}_n^\sigma}(\underline{a}\,I + \underline{b}\,\underline{S}\,)\underline{P}_n^\sigma \to \underline{a}\,I + \underline{b}\,\underline{S}$ *strongly.*

Now let us consider A_n^*. We have $(\widetilde{\underline{L}_n^\sigma}\underline{a}\,\underline{P}_n^\sigma)^* = \widetilde{\underline{L}_n^\sigma}\underline{a}^*\underline{P}_n^\sigma$. Indeed,

$$\langle\widetilde{\underline{L}_n^\sigma}\underline{a}\,\underline{P}_n^\sigma\underline{u}\,,\underline{v}\,\rangle_{\sigma,k} = \sum_{i=1}^k\sum_{j=1}^k\langle\widetilde{L_n^\sigma}a_{ij}P_n^\sigma u_j, v_i\rangle_\sigma = \sum_{i=1}^k\sum_{j=1}^k\langle u_j, \widetilde{L_n^\sigma}\overline{a_{ij}}P_n^\sigma v_i\rangle_\sigma$$

$$= \langle\underline{u}\,,\widetilde{\underline{L}_n^\sigma}\underline{a}^*\underline{P}_n^\sigma\underline{v}\,\rangle_{\sigma,k}.$$

Since $\widetilde{a}$, w are scalar valued functions, one can easily obtain the same expression for the operator $(\widetilde{\underline{L}_n^\sigma}(\widetilde{a}I + i\underline{S}\,)w\underline{P}_n^\sigma)^*$ as before and can prove its strong convergence. To investigate $\widetilde{A_n}$, we first remark that for $\underline{u} \in (\mathbf{L}_\sigma^2)^k$, $\underline{a} \in \mathbf{R}^{k\times k}$ we have due to Lemma 4.19

$$\underline{W}_n^\sigma\widetilde{\underline{L}_n^\sigma}\underline{a}\,\underline{W}_n^\sigma\underline{u} = \left(W_n^\sigma\widetilde{L_n^\sigma}\sum_{j=1}^k a_{ij}W_n^\sigma u_j\right)_{i=1}^k = \left(\sum_{j=1}^k\widetilde{L_n^\sigma}a_{ij}P_n^\sigma u_j\right)_{i=1}^k = \widetilde{\underline{L}_n^\sigma}\underline{a}\,\underline{P}_n^\sigma\underline{u}\,.$$

Lemma 6.3 *If $\underline{b} \in (\mathbf{PC}_0)^{k\times k}$, then the sequence $\underline{W}_n^\sigma \widetilde{\underline{L}_n^\sigma} \underline{b}\,\underline{S}\,\underline{W}_n^\sigma$ is strongly convergent to $-\underline{b}\,w_\sigma^{-1}\underline{S}\,w_\sigma I$.*

Proof. Let first $\underline{b} \in (\mathbf{C}_{0,0}^{1,\eta})^{k\times k}$, $\eta > \frac{1+\max\{\alpha,\beta,0\}}{2}$. Then a decomposition analogous to (4.5) remains valid, where w_σ and φ have to be understood as scalar-valued. The summand K_1 occurring there contains entries of the form $(b_{ij}S - Sb_{ij}I)_{i,j=1}^k$, all of them being elements of $\mathcal{K}(\mathbf{L}_\sigma^2, \mathbf{C}^{0,\lambda})$, and is therefore itself a compact operator from $(\mathbf{L}_\sigma^2)^k$ to some space $(\mathbf{C}^{0,\lambda})^k$, analogous relations hold for K_2. The following steps can be taken literally from the proof of Proposition 4.22. In particular, we approximate a function $\underline{b} \in (\mathbf{PC}_0)^{k\times k}$ with a finite number of jumps by sums of the form $\sum_{j=1}^m \chi_j \widetilde{\underline{b}}_j$, where χ_j are scalar-valued characteristic functions and $\widetilde{\underline{b}}_j \in (\mathbf{C}_{0,0}^{1,\eta})^{k\times k}$. ■

The arguments concerning $\widetilde{A_n}^*$ are completely analogous.

6.3 Local theory

The algebra $\underline{\mathcal{A}}$ and the ideal $\underline{\mathcal{I}}$ (and the corresponding objects related to the unit circle) are defined analogously to the scalar case. For $\tau \in [-1,1]$, let $\underline{M}_\tau = \left\{\{\widetilde{\underline{L}_n^\sigma}(\delta_{ij}f)_{i,j=1}^k \underline{P}_n^\sigma\} + \underline{\mathcal{I}} : f \in m_\tau\right\}$. (Note that the functions f in this definition are scalar-valued.) Clearly, $\{\underline{M}_\tau\}_{\tau\in[-1,1]}$ is a covering system of localizing classes in $\underline{\mathcal{A}}/\underline{\mathcal{I}}$.

If we define $\underline{F} := (\delta_{ij}F)_{i,j=1}^k : (\mathbf{L}_\sigma^2)^k \to (L^2(\mathrm{T}))^k$, we can easily verify that the identities given in Lemma 5.3 remain to hold in the matrix case. Hence, Proposition 5.7 and Lemma 5.10 hold, too, and we are able to show the commutativity of the cosets $\{\widetilde{\underline{L}_n^\sigma \underline{a}\,I + \underline{b}\,\underline{S}}\,\underline{P}_n^\sigma\} + \underline{\mathcal{I}}$ ($\underline{a} \in \mathbf{PC}^{k\times k}$, $\underline{b} \in \mathbf{PC}_0^{k\times k}$) with all elements of $\bigcup_{\tau\in[-1,1]} \underline{M}_\tau$. (Note that the functions $f \in m_\tau$ occurring in the localizing classes commute with matrix-valued functions since they are scalar-valued.) In the same way as above we can now show the result about the $\underline{M}_\tau$-equivalence:

Lemma 6.4 *Let $\tau \in [-1,1]$, $\underline{a}, \underline{a}_\tau \in \mathbf{PC}^{k\times k}$, $\underline{b}, \underline{b}_\tau \in (\mathbf{PC}_0)^{k\times k}$ such that*

$$\underline{a}_\tau(\tau \pm 0) = \underline{a}(\tau \pm 0), \qquad \underline{b}_\tau(\tau \pm 0) = \underline{b}(\tau \pm 0).$$

Then

$$\{\widetilde{\underline{L}_n^\sigma}(\underline{a}\,I + \underline{b}\,\underline{S})\underline{P}_n^\sigma\} + \underline{\mathcal{I}} \overset{\underline{M}_\tau}{\sim} \{\widetilde{\underline{L}_n^\sigma}(\underline{a}_\tau I + \underline{b}_\tau \underline{S})\underline{P}_n^\sigma\} + \underline{\mathcal{I}}.$$

Lemma 6.5 (cf. the proof of [31], Th. 3.1) *Assume that $\underline{a}, \underline{b} \in \mathbf{PC}^{k\times k}(\mathrm{T})$ and that $A^{\mathbf{T}} = \underline{a}\,I + \underline{b}\,\underline{S}_{\mathbf{T}}$ is a Fredholm operator in $(\mathbf{L}^2(\mathrm{T}))^k$. Then the coset $\{\underline{M}_n^{\mathbf{T}} A^{\mathbf{T}} \underline{P}_n^{\mathbf{T}}\} + \underline{\mathcal{I}}^{\mathbf{T}}$ is invertible in $\underline{\mathcal{A}}^{\mathbf{T}}/\underline{\mathcal{I}}^{\mathbf{T}}$.*

At this point we have to give a Lemma on the Fredholmness of singular integral operators with matrix-valued coefficients.

Lemma 6.6 ([39], Th. 6.1) *Let $a, b \in \mathbf{PC}^{k\times k}[-1,1]$. Then the operator $\underline{a}\,I + \underline{b}\,\underline{S}$ is Fredholm in $(\mathbf{L}_\sigma^2)^k$ if and only if*

$$\inf_{x\in[-1,1]} |\det(\underline{a}(x) \pm \underline{b}(x))| > 0$$

and $\mathbf{c}(x,\mu) \neq 0$ *for all* $x \in [-1,1]$, $\mu \in [0,1]$, *where* $\mathbf{c}$ *is defined analogously to (5.8), but with* $c(x) = \det([\underline{a}(x) - \underline{b}(x)]^{-1}[\underline{a}(x) + \underline{b}(x)])$. *An analogous relation holds for singular integral operators in* $(\mathbf{L}^2(\mathbb{T}))^k$ *with respect to the symbol* $\mathbf{c}^{\mathbf{T}}$ *(compare Definition 5.13).*

Theorem 6.7 (cf. [33], Th. 7.2) *Let* $\underline{a} \in \mathbf{PC}^{k\times k}[-1,1]$, $\underline{b} \in \mathbf{PC}_0^{k\times k}[-1,1]$, $A = \underline{a}\,I + \underline{b}\,\underline{S}$. *Then the sequence* $\{A_n\} = \{\widetilde{\underline{L}_n^\sigma A \underline{P}_n^\sigma}\}$ *is stable if and only if the operators* A *and* $\widetilde{A} = \underline{a}\,I - \underline{b}\,w_\sigma^{-1}\underline{S}\,w_\sigma I$ *are invertible in* $(\mathbf{L}^2_\sigma)^k$.

Proof. It remains to consider the invertibility of the coset $\{A_n\} + \underline{\mathcal{I}}$. Since $\underline{b}(\pm 1) = 0$, Lemma 6.6 shows that the invertibility of A and $\widetilde{A}$ implies that $A^{\mathbf{T}} = \widehat{\underline{a}} - \psi\widehat{\underline{b}}\,\underline{S}_{\mathbf{T}}$ is Fredholm in $(\mathbf{L}^2(\mathbb{T}))^k$. Hence, in view of Lemma 6.5, the coset $\{\underline{M}_n^{\mathbf{T}} A^{\mathbf{T}} \underline{P}_n^{\mathbf{T}}\} + \underline{\mathcal{I}}^{\mathbf{T}}$ is invertible. In the same way as in the proof of Theorem 5.15 one can now conclude the invertibility of $\{A_n\} + \underline{\mathcal{I}}$. ∎

Remark 6.8 *In contrast to the scalar case, we must explicitely require that* $\widetilde{A}$ *is invertible. The invertibility of* A *implies that* $\widetilde{A}$ *is a Fredholm operator with index 0, but in the system case this does not necessarily imply invertibility.*

7 Sobolev-like spaces

In this section we introduce a scale of subspaces of $\mathbf{L}^2_\sigma$ analogous to that one considered in [4]. We will show some mapping properties of the operators we investigate in these spaces. That will enable us to prove optimal convergence rates for the Galerkin and collocation methods and to construct certain fast algorithms for the efficient numerical solution of singular integral equations perturbed by an integral operator with a smooth kernel, being based on the mapping properties of the operators involved.

7.1 Definition and basic properties of the spaces $\widetilde{\mathbf{L}}^2_{\sigma,s}$

The following considerations are valid for the case of an arbitrary Jacobi weight v the exponents of which satisfy (1.4). In [4], numerical methods being based on algebraic polynomials were investigated in a scale of Sobolev-like spaces $\mathbf{L}^2_{v,s}$, $s \geq 0$, defined in the following way:

$$\mathbf{L}^2_{v,s} = \left\{ u \in \mathbf{L}^2_v : \|u\|_{v,s} := \left(\sum_{n=0}^{\infty} (1+n)^{2s} |\langle u, p_n^v \rangle_v|^2 \right)^{1/2} < \infty \right\}.$$

Thus, it seems appropriate to introduce analogous spaces associated with the orthonormal system $\{\widetilde{u}_n\}$ for the investigation of our numerical methods.

Definition 7.1 *For $s \geq 0$ let*

$$\widetilde{\mathbf{L}}^2_{\sigma,s} = \left\{ u \in \mathbf{L}^2_\sigma : \|u\|_{\sigma,s,\sim} := \left(\sum_{n=0}^{\infty} (1+n)^{2s} |\langle u, \widetilde{u}_n \rangle_\sigma|^2 \right)^{1/2} < \infty \right\}.$$

It is clear that hereby also an inner product in $\widetilde{\mathbf{L}}^2_{\sigma,s}$ is given by

$$\langle u, w \rangle_{\sigma,s} := \sum_{n=0}^{\infty} (1+n)^{2s} \langle u, \widetilde{u}_n \rangle_\sigma \overline{\langle w, \widetilde{u}_n \rangle_\sigma}.$$

(For the sake of brevity we will in the following omit the reference to σ and write $\|\cdot\|_{s,\sim}$ instead of $\|\cdot\|_{\sigma,s,\sim}$.)

Lemma 7.2 *The multiplication operator $w_{v,\sigma^{-1}} I$ is an isometric isomorphism from $\mathbf{L}^2_{v,s}$ onto $\widetilde{\mathbf{L}}^2_{\sigma,s}$ that transforms p_n^v into $\widetilde{u}_n$ $(n = 0, 1, 2, \ldots)$.*

Thus, many properties of the spaces $\widetilde{\mathbf{L}}^2_{\sigma,s}$ can be easily derived from those of $\mathbf{L}^2_{v,s}$ shown, for instance, in [4]. In the following lemma, we will summarize some basic properties we will make use of in the sequel. We use the notation $Q_n^\sigma = I - P_n^\sigma$.

Lemma 7.3 (cf. [4], Lemmas 2.1, 2.2) *(i) $\|u\|_{t,\sim} \leq \|u\|_{s,\sim}$ for $0 \leq t \leq s$*

(ii) $\widetilde{\mathbf{L}}^2_{\sigma,s}$ is a Hilbert space.

(iii) Q_n^σ converges strongly to 0 in $\widetilde{\mathbf{L}}^2_{\sigma,s}$.

(iv) $\|Q_n^\sigma u\|_{t,\sim} \le (1+n)^{t-s}\|u\|_{s,\sim}$ *for* $t \le s$

(v) $\|P_n^\sigma u\|_{t,\sim} \le \begin{cases} \|u\|_{s,\sim}, & t < s \\ n^{t-s}\|u\|_{s,\sim}, & t \ge s \end{cases}$

Corollary 7.4 *Let $A \in \mathcal{L}(\mathbf{L}^2_\sigma)$ and consider the Galerkin method with the approximate equations (2.4) for the approximate solution of the operator equation*

$$Au = f, \qquad f \in \mathbf{L}^2_\sigma. \tag{7.1}$$

Assume that the solution u of the latter equation belongs to $\widetilde{\mathbf{L}}^2_{\sigma,s}$ and that the sequence $\{A_{n,P}\} = \{P_n^\sigma A P_n^\sigma\}$ is stable in $\mathbf{L}^2_\sigma$. Then we have the convergence rate

$$\|u - u_n\|_{t,\sim} \le \text{const}\, n^{t-s}\|u\|_{s,\sim} \qquad \textit{for all} \quad 0 \le t \le s.$$

Proof. Since $\{P_n^\sigma A P_n^\sigma\}$ is stable, we have

$$\begin{aligned} \|P_n^\sigma u - u_n\|_{t,\sim} &\le n^t \|P_n^\sigma u - u_n\|_\sigma \\ &\le \text{const}\, n^t \|A_{n,P}P_n^\sigma u - P_n^\sigma f\|_\sigma = \text{const}\, n^t \|P_n^\sigma A(P_n^\sigma u - u)\|_\sigma \\ &\le \text{const}\, n^t \|A\|_{\mathcal{L}(\mathbf{L}^2_\sigma)} (1+n)^{-s}\|u\|_{s,\sim}. \end{aligned}$$

and the corollary is proved (note that $\|Q_n^\sigma u\|_{t,\sim} \le \text{const}\, n^{t-s}\|u\|_{s,\sim}$). ■

The fact that P_n^σ converges strongly to I implies that $\text{span}\{\widetilde{u}_n\}_{n=0}^\infty$ is dense in $\widetilde{\mathbf{L}}^2_{\sigma,s}$ for all $s \ge 0$.

Furthermore, the embedding operator from $\widetilde{\mathbf{L}}^2_{\sigma,s}$ into $\widetilde{\mathbf{L}}^2_{\sigma,t}$ is compact for all $s > t$, since the finite-dimensional operators P_n^σ converge uniformly to it. Thus, we have

Lemma 7.5 (cf. [4]) *For $s > t$, the space $\widetilde{\mathbf{L}}^2_{\sigma,s}$ is densely and compactly embedded in the space $\widetilde{\mathbf{L}}^2_{\sigma,t}$.*

In the following subsections we will sometimes make use of an interpolation theorem to verify the boundedness of a linear operator in Sobolev spaces. The preceding two lemmas show that the spaces $\widetilde{\mathbf{L}}^2_{\sigma,s}$, $s \ge 0$, satisfy the assumptions of this theorem:

Theorem 7.6 ([44], 1.32; [3]) *Let $\{H^s\}_{s\ge 0}$ be a scale of Hilbert spaces (the norms of which are denoted by $\|\cdot\|_s$) such that $\|u\|_t \le \|u\|_s$ and H^s is dense in H^t for $t < s$. Let furthermore $A \in \mathcal{L}(H^{s_1}) \cap \mathcal{L}(H^{s_2})$, $s_1 < s_2$. Then we have $A \in \mathcal{L}(H^s)$ and*

$$\|A\|_{\mathcal{L}(H^s)} \le \|A\|^\lambda_{\mathcal{L}(H^{s_1})}\|A\|^{1-\lambda}_{\mathcal{L}(H^{s_2})}$$

for all $s \in [s_1, s_2]$, where $s = \lambda s_1 + (1-\lambda)s_2$.

7.2 Embedding theorems, equivalent norms, and interpolation operators in $\widetilde{\mathbf{L}}^2_{\sigma,s}$

We are going to summarize some results about the embedding of $\widetilde{\mathbf{L}}^2_{\sigma,s}$ in spaces of continuous (or weighted continuous) functions. This shows that error estimates in the norms $\|\cdot\|_{s,\sim}$ under certain conditions also provide (weighted) uniform estimates.

Definition 7.7 *Let μ be a Jacobi weight. By $\mathbf{C}_\mu$ we denote the Banach space of all functions $u : (-1,1) \to \mathbb{C}$ for which μu can be extended continuously to the closed interval $[-1,1]$. The norm in this space is given by $\|u\|_{\mathbf{C}_\mu} = \|\mu u\|_\infty$.*

Lemma 7.8 ([2], Lemma 1.3) *For $v = v^{\gamma,\delta}$, $n = 1,2,\ldots$, and $x \in [-1,1]$ we have the relation*

$$|p_n^v(x)| \le \text{const} \left(\sqrt{1-x} + \frac{1}{n}\right)^{-\gamma-\frac{1}{2}} \left(\sqrt{1+x} + \frac{1}{n}\right)^{-\delta-\frac{1}{2}}.$$

The following proposition is a slight generalization of [40, Theorem 7].

Proposition 7.9 (cf. [40], Theorem 7) *Let $s > \frac{1}{2}$. Then $\mathbf{L}^2_{v,s}$ is compactly embedded in $\mathbf{C}_\mu$, where $\mu(x) = (1-x)^{\max\{1/4+\gamma/2,0\}}(1+x)^{\max\{1/4+\delta/2,0\}}$.*

Proof. Let $u \in \mathbf{L}^2_{v,s}$, $\widetilde{u} = \mu u$, $\widetilde{v} = \mu^{-2} v$, $\widetilde{p_n^v} = \mu p_n^v$. Then we have

$$\widetilde{u} = \mu \sum_{n=0}^\infty \langle u, p_n^v\rangle_v p_n^v = \sum_{n=0}^\infty \langle \widetilde{u}, \widetilde{p_n^v}\rangle_{\widetilde{v}} \widetilde{p_n^v}.$$

According to Lemma 7.8, the estimation

$$|\widetilde{p_n^v}(x)| \le \text{const} \left(\sqrt{1-x} + \frac{1}{n}\right)^{-\gamma-\frac{1}{2}} (1-x)^{\max\{\frac{1}{4}+\frac{\gamma}{2},0\}}$$

$$\times \left(\sqrt{1+x} + \frac{1}{n}\right)^{-\delta-\frac{1}{2}} (1+x)^{\max\{\frac{1}{4}+\frac{\delta}{2},0\}} \le \text{const}$$

holds, which means $|\langle \widetilde{u}, \widetilde{p_n^v}\rangle_{\widetilde{v}} \widetilde{p_n^v}| \le \text{const}\, |\langle \widetilde{u}, \widetilde{p_n^v}\rangle_{\widetilde{v}}|$. Further, we have

$$\sum_{n=0}^\infty |\langle \widetilde{u}, \widetilde{p_n^v}\rangle_{\widetilde{v}}| \le \left(\sum_{n=0}^\infty (1+n)^{2s} |\langle u, p_n^v\rangle_v|^2\right)^{\frac{1}{2}} \left(\sum_{n=0}^\infty (1+n)^{-2s}\right)^{\frac{1}{2}}$$

$$\le \text{const}\, \|u\|_{v,s},$$

from which we conclude that the Fourier series of $\widetilde{u}$ with respect to the system $\{\widetilde{p_n^v}\}$ converges uniformly and therefore has a continuous sum. At the same time, this estimate shows the continuity of the embedding and hence, in view of Lemma 7.5, also its compactness. ■

Corollary 7.10 *If $s > \frac{1}{2}$, the space $\widetilde{\mathbf{L}}^2_{\sigma,s}$ is compactly embedded in $\mathbf{C}_w$, where $w(x) = \mu(x) w^{-1}_{v,\sigma^{-1}}(x) = (1-x)^{\alpha/2+\max\{-\gamma/2,1/4\}}(1+x)^{\beta/2+\max\{-\delta/2,1/4\}}$.*

Remark 7.11 *We only mention that it is possible to get more subtle results on the behaviour of functions from $\widetilde{\mathbf{L}}^2_{\sigma,s}$. Let us only consider the example $v = \varphi$, $\sigma = \varphi^{-1}$. Then we have $w \equiv 1$ in the preceding corollary. Because of the continuous embedding $\widetilde{\mathbf{L}}^2_{\sigma,s} \subset \mathbf{C}$ in this case we obtain the conclusion that $u \in \widetilde{\mathbf{L}}^2_{\sigma,s}$ can be approximated uniformly by elements of* span $\{\varphi U_n\}$, *which leads in particular to the relation $u(\pm 1) = 0$. We will dispense with further considerations in this direction, however.*

Remark 7.12 *The preceding assertions have to be understood in the following sense: Under the assumption $s > \frac{1}{2}$, for every class $u \in \widetilde{\mathbf{L}}^2_{\sigma,s}$ there exists a representative of u that belongs to a space of weighted continuous functions.*

Proposition 7.13 ([4], Theorem 2.8) *Let $s > \frac{1}{2} + \max\{\gamma + \frac{1}{2}, \delta + \frac{1}{2}, 0\}$. Then $\mathbf{L}^2_{v,s}$ is compactly embedded in $\mathbf{C}[-1,1]$.*

Corollary 7.14 *For $s > \frac{1}{2} + \max\{\gamma + \frac{1}{2}, \delta + \frac{1}{2}, 0\}$ we have the compact embedding $\widetilde{\mathbf{L}}^2_{\sigma,s} \subset \mathbf{C}_{w^{-1}_{v,\sigma^{-1}}}$, in particular $\widetilde{\mathbf{L}}^2_{\sigma,s} \subset \mathbf{C}[-1,1]$ if $\alpha \le \gamma, \beta \le \delta$.*

The following property will only be given with respect to $\mathbf{L}^2_{v,s}$, the corresponding assertions for $\widetilde{\mathbf{L}}^2_{\sigma,s}$ can be easily derived from Lemma 7.2.

Proposition 7.15 ([4], Th. 2.9 and 2.13) *Let $0 < \eta \le 1$, $s > \frac{1}{2} + \eta$. Then $\mathbf{L}^2_{v,s}$ is compactly embedded in $\mathbf{C}^{0,\eta}[-1+\varepsilon, 1-\varepsilon]$ for arbitrary $0 < \varepsilon < 1$. If r is a positive integer, $s > \frac{1}{2} + r$, then $\mathbf{L}^2_{v,s}$ is compactly embedded in $\mathbf{C}^r[-1+\varepsilon, 1-\varepsilon]$ for arbitrary $0 < \varepsilon < 1$.*

The preceding propositions show that functions from $\widetilde{\mathbf{L}}^2_{\sigma,s}$ are continuous at least on $(-1,1)$ if $s > \frac{1}{2}$. Hence, it makes sense to apply interpolation operators to such functions.

Theorem 7.16 ([38], Theorem 3.4) *Let $s > \frac{1}{2}$. Then the Lagrange interpolation operator L^v_n is bounded on $\mathbf{L}^2_{v,s}$, and the estimation*

$$\|f - L^v_n f\|_{v,t} \le \text{const}\, n^{t-s} \|f\|_{v,s}$$

holds for all $0 \le t \le s$, where the constant does not depend on t.

Corollary 7.17 *For $s > \frac{1}{2}$, we have $\widetilde{L^\sigma_n} \in \mathcal{L}(\widetilde{\mathbf{L}}^2_{\sigma,s})$ and the estimation*

$$\left\| f - \widetilde{L^\sigma_n} f \right\|_{s,\sim} \le \text{const}\, n^{t-s} \|f\|_{s,\sim} \qquad \text{for all} \quad 0 \le t \le s.$$

Remark 7.18 *The preceding estimations show that $\widetilde{L^\sigma_n}$ converges uniformly to the embedding operator from $\widetilde{\mathbf{L}}^2_{\sigma,s}$ into $\widetilde{\mathbf{L}}^2_{\sigma,t}$ if $s > \frac{1}{2}$ and $0 \le t < s$. In $\widetilde{\mathbf{L}}^2_{\sigma,s}$ we still have strong convergence to the identity operator, since $\widetilde{L^\sigma_n}$ is uniformly bounded and converges to I on* span$\{\widetilde{u}_n\}_{n=0}^\infty$.

Having these properties of the interpolation operators $\widetilde{L_n^\sigma}$ at our disposal, we are now able to give an error estimate for an abstract collocation method. This is not restricted to the singular integral operators we are investigating.

Let $A \in \mathcal{L}(\mathbf{L}^2_\sigma)$ be such that the operators $A_{n,L} = \widetilde{L_n^\sigma} A P_n^\sigma$ are in $\mathcal{L}(\mathbf{L}^2_\sigma)$. Further, consider the operator equation (7.1), where the right hand side f is such that $\left\| \widetilde{L_n^\sigma} f - f \right\|_\sigma$ converges to 0 for $n \to \infty$. Then it makes sense to use a collocation method of the type (2.6) to solve (7.1) approximately.

Theorem 7.19 ([33], Th. 8.3, Rem. 8.4) *Assume that $A \in \mathcal{L}(\widetilde{\mathbf{L}}^2_{\sigma,r})$ for some r, $\frac{1}{2} < r \le s$. If the sequence $\{A_{n,L}\} = \{\widetilde{L_n^\sigma} A P_n^\sigma\}$ is stable in $\mathbf{L}^2_\sigma$ and if the solution u of equation (7.1) belongs to $\widetilde{\mathbf{L}}^2_{\sigma,s}$, then*

$$\|u_n - u\|_{t,\sim} \le \text{const}\, n^{t-s} \|u\|_{s,\sim}\,, \quad 0 \le t \le s\,,$$

where $u_n \in \operatorname{im} P_n^\sigma$ is the solution of (2.6).

Proof. Since $\{A_{n,L}\}$ is assumed to be stable we have, in view of Lemma 7.3 (iv),(v),

$$\begin{aligned}
\|P_n^\sigma u - u_n\|_{t,\sim} &\le n^t \|P_n^\sigma u - u_n\|_\sigma \\
&\le \text{const}\, n^t \left\| A_{n,L} P_n^\sigma u - \widetilde{L_n^\sigma} f \right\|_\sigma = \text{const}\, n^t \left\| \widetilde{L_n^\sigma} A (P_n^\sigma u - u) \right\|_\sigma \\
&\le \text{const}\, n^t \left(\left\| (\widetilde{L_n^\sigma} - I) A (P_n^\sigma u - u) \right\|_\sigma + \|A(P_n^\sigma u - u)\|_\sigma \right) \\
&\le \text{const}\, n^t \left(n^{-r} \|A\|_{\mathcal{L}(\widetilde{\mathbf{L}}^2_{\sigma,r})} (1+n)^{r-s} \|u\|_{s,\sim} + \|A\|_{\mathcal{L}(\mathbf{L}^2_\sigma)} (1+n)^{-s} \|u\|_{s,\sim} \right),
\end{aligned}$$

and the theorem is proved (note that $\|Q_n^\sigma u\|_{t,\sim} \le \text{const}\, n^{t-s} \|u\|_{s,\sim}$). ∎

Using the definition of $\widetilde{\mathbf{L}}^2_{\sigma,s}$, it is in most cases difficult to decide whether a given function belongs to such a space, since this requires the knowledge of its Fourier coefficients $\langle u, \widetilde{u}_n \rangle_\sigma$, or at least a sharp upper estimate for them. The following relations provide a simple criterion for a function to belong to $\widetilde{\mathbf{L}}^2_{\sigma,s}$ in the case that s is an integer.

Theorem 7.20 (cf. [4], Th. 2.17) *Let $D = \frac{d}{dx}$ be the operator of differentiation (in the generalized sense), and let s be a positive integer. Then u belongs to $\mathbf{L}^2_{v,s}$ if and only if $\varphi^k D^k u \in \mathbf{L}^2_v$, $k = 0, \ldots, s$, and the following norm equivalence holds:*

$$\|u\|_{v,s} \sim \sum_{k=0}^{s} \left\| \varphi^k D^k u \right\|_v .$$

Corollary 7.21 *If $s \ge 0$ is an integer, then $u \in \widetilde{\mathbf{L}}^2_{\sigma,s}$ if and only if $\varphi^k D^k (w^{-1}_{v,\sigma^{-1}} u) \in \mathbf{L}^2_v$, $k = 0, \ldots, s$, and we have $\|u\|_{s,\sim} \sim \sum_{k=0}^{s} \left\| \varphi^k D^k (w^{-1}_{v,\sigma^{-1}} u) \right\|_v$.*

It follows a further characterization of $\widetilde{\mathbf{L}}^2_{\sigma,s}$ that will be of use when investigating the mapping properties of certain operators in such spaces.

Lemma 7.22 (cf. [37], Prop. 2.22, Rem. 2.23) *For $u \in \mathbf{L}^2_\sigma$ we define*

$$\widetilde{E_n}(u) := \inf_{u_n \in X_n} \|u - u_n\|_\sigma, \quad n = 1, 2, \ldots, \quad \widetilde{E_0}(u) := \|u\|_\sigma .$$

Then $u \in \widetilde{\mathbf{L}}^2_{\sigma,s}$ holds if and only if

$$\left(\sum_{n=0}^{\infty} (1+n)^{2s-1} \left[\widetilde{E_n}(u) \right]^2 \right)^{\frac{1}{2}} < \infty,$$

and the latter expression is an equivalent norm in $\widetilde{\mathbf{L}}^2_{\sigma,s}$.

Proof. We have

$$\sum_{n=0}^{\infty} (1+n)^{2s-1} \left[\widetilde{E_n}(u) \right]^2 = \sum_{n=0}^{\infty} (1+n)^{2s-1} \sum_{k=n}^{\infty} |\langle u, \widetilde{u}_k \rangle_\sigma|^2$$

$$= \sum_{k=0}^{\infty} |\langle u, \widetilde{u}_k \rangle_\sigma|^2 \sum_{n=0}^{k} (1+n)^{2s-1}.$$

Further, one can show that there exist constants $c_1, c_2 > 0$ such that

$$c_1 (1+k)^{2s} \le \sum_{n=0}^{k} (1+n)^{2s-1} \le c_2 (1+k)^{2s}$$

by interpreting the sum as an integral sum for $\int_0^{k+1} (1+x)^{2s-1}\, dx$. ∎

7.3 Mapping properties of singular integral operators in spaces $\widetilde{\mathbf{L}}^2_{\sigma,s}$

Now we are going to deal with the boundedness of certain operators in $\widetilde{\mathbf{L}}^2_{\sigma,s}$. We begin with multiplication operators. First of all, there is a simple criterion for the case of an integer s. In the case of non-integer s we rely on the interpolation theorem 7.6.

Lemma 7.23 (cf. [25], Lemma 3.5) *Let $s \ge 0$ be an integer, and let a be a function defined on $(-1, 1)$ that satisfies $\varphi^k D^k a \in \mathbf{L}^\infty$ for $k = 0, \ldots, s$. Then the multiplication operator aI is bounded on $\widetilde{\mathbf{L}}^2_{\sigma,s}$, and its norm can be estimated by*

$$\|aI\|_{\mathcal{L}(\widetilde{\mathbf{L}}^2_{\sigma,s})} \le \text{const} \sum_{k=0}^{s} \left\| \varphi^k D^k a \right\|_\infty =: \text{const} \, \|a\|_{\infty,\varphi,s} .$$

Here the derivatives have to be understood in the sense of generalized differentiation.

Proof. First recall that $\|au\|_{s,\sim} = \left\|aw_{v,\sigma^{-1}}^{-1}u\right\|_{v,s}$ and $\|u\|_{s,\sim} = \left\|w_{v,\sigma^{-1}}^{-1}u\right\|_{v,s}$, hence $aI \in \mathcal{L}(\widetilde{\mathbf{L}}^2_{\sigma,s})$ if and only if $aI \in \mathcal{L}(\mathbf{L}^2_{v,s})$. We show the latter relation by induction. For $s = 0$ it is obvious. Let the relation be true for some s. Now consider the assertion with respect to $s+1$ and keep in mind the norm equivalence from Theorem 7.20. We have

$$\begin{aligned}
\|au\|_{v,s+1} &\le \text{const} \sum_{k=0}^{s+1} \left\| \varphi^k \sum_{j=0}^{k} \binom{k}{j} D^{k-j}a\, D^j u \right\|_v \\
&\le \text{const} \left(\|a\|_{\infty,\varphi,s} \|u\|_{v,s} + \sum_{j=0}^{s+1} \binom{s+1}{j} \left\|\varphi^{s+1-j} D^{s+1-j} a\right\|_\infty \left\|\varphi^j D^j u\right\|_v \right) \\
&\le \text{const} \|a\|_{\infty,\varphi,s+1} \|u\|_{v,s+1} ,
\end{aligned}$$

and the proof is complete. ■

The following proposition gives a more subtle condition for the boundedness of multiplication operators.

Proposition 7.24 (U. Luther, unpublished) *Let $s > \frac{1}{2}$, $\varrho = v^{r,t}$, where $r = \min\{\gamma, -\frac{1}{2}\}$, $t = \min\{\delta, -\frac{1}{2}\}$. If $a \in \mathbf{L}^2_{\varrho,s}$, then $aI \in \mathcal{L}(\mathbf{L}^2_{v,s})$ and $\|aI\|_{\mathcal{L}(\mathbf{L}^2_{v,s})} \le$ const $\|a\|_{\varrho,s}$.*

Proof. First we note the norm equivalence

$$\|u\|^2_{v,s} \sim \sum_{n=0}^{\infty} (1+n)^{2s-1} [E_{v,n}(u)]^2,$$

where $E_{v,n}(u) = \inf_{p\in\Pi_n} \|u-p\|_v$ and Π_n denotes the set of all polynomials of degree less than n (compare Lemma 7.22 and its proof). Denote by P_n the orthogonal projection from $\mathbf{L}^2_v$ onto Π_n, that is, $\|u - P_n u\|_v = E_{v,n}(u)$.

Now let n be a fixed integer and define $k(n) = l(n) = \frac{n}{2}$ if n is even and $k(n) = \frac{n+1}{2}$, $l(n) = \frac{n-1}{2}$ otherwise. We have

$$\begin{aligned}
E_{v,n}(au) &\le \left\|au - (P_{l(n)}a)(P_{k(n)}u)\right\|_v \\
&\le \left\|(a - P_{l(n)}a)\, u\right\|_v + \left\|(P_{l(n)}a)\,(u - P_{k(n)}u)\right\|_v \\
&\le \|u\|_{\mathbf{C}_\mu} \left\|a - P_{l(n)}a\right\|_\varrho + \left\|P_{l(n)}a\right\|_\infty \left\|u - P_{k(n)}u\right\|_v ,
\end{aligned}$$

where μ is the same as in Proposition 7.9. The continuous embedding stated in the latter proposition allows us to estimate the above terms further by

$$\text{const} \left(\|u\|_{v,s} E_{\varrho,l(n)}(a) + \|a\|_{\varrho,s} E_{v,k(n)}(u) \right).$$

(Note that $\left\|P_{l(n)}a\right\|_\infty \le \text{const}\ \left\|P_{l(n)}a\right\|_{\varrho,s} \le \text{const}\ \|a\|_{\varrho,s}$ due to Proposition 7.9 and [4, Lemma 2.2].) Hence, we have

$$\|au\|^2_{v,s} \le \text{const}\ \left(\|u\|^2_{v,s} \sum_{n=0}^\infty (1+n)^{2s-1}[E_{\varrho,l(n)}(a)]^2 \right.$$

$$\left. + \|a\|^2_{\varrho,s} \sum_{n=0}^\infty (1+n)^{2s-1}[E_{v,k(n)}(u)]^2 \right) \le \text{const}\ \|a\|^2_{\varrho,s} \|u\|^2_{v,s},$$

and we are done. (Note that $(1+n)^{2s-1} \le \text{const}\ (1+l(n))^{2s-1}$.) ■

Corollary 7.25 *Let $s > \frac{1}{2}$. Then $a \in \mathbf{L}^2_{\varrho,s}$ implies that $aI \in \mathcal{L}(\widetilde{\mathbf{L}}^2_{\sigma,s})$, where ϱ is defined as in the preceding proposition. If in particular s is an integer, the condition $\varphi^k D^k a \in \mathbf{L}^2_\varrho$, $k = 0, \ldots, s$, is sufficient for $aI \in \mathcal{L}(\widetilde{\mathbf{L}}^2_{\sigma,s})$.*

To investigate the mapping properties of the Cauchy singular integral, we confine ourselves for technical reasons to considering the case $v = \varphi$. We treat the operator bS as a unit and make use of the decomposition (4.5).

First consider the term $b\varphi^{-1}(eI - V^*)$. It is obvious that $V^* \in \mathcal{L}(\widetilde{\mathbf{L}}^2_{\sigma,s})$ for all $s \ge 0$, the same applies to the operator eI due to Corollary 7.25. Hence, we concentrate on the multiplication operator $b\varphi^{-1}I$.

Lemma 7.26 *Let $s \ge 1$ be an integer and assume that $D^s b$ belongs to $\mathbf{L}^2$ on every compact subinterval of $(-1,1)$ and*

$$(D^j b)(x) \le \text{const}\ (1-x^2)^{\frac{s+1/2}{2} - j + \varepsilon}, \quad j = 0, \ldots, s,$$

for some $\varepsilon > 0$ in a neighbourhood of 1 and -1. Then $b\varphi^{-1}I \in \mathcal{L}(\widetilde{\mathbf{L}}^2_{\sigma,s})$.

Proof. According to Corollary 7.25 we verify the conditions

$$\varphi^k D^k (b\varphi^{-1}) \in \mathbf{L}^2_{\varphi^{-1}}, \qquad k = 0, \ldots, s. \tag{7.2}$$

We have

$$D^k(b\varphi^{-1}) = \sum_{j=0}^k \binom{k}{j} (D^j b)\,(D^{k-j}(\varphi^{-1}))$$

as well as

$$(D^{k-j}\varphi^{-1})(x) = \sum_{l=0}^{k-j} c_{lj}(1+x)^{-\frac{1}{2}-l}(1-x)^{-\frac{1}{2}-(k-j)+l} = O\left((1-x^2)^{-\frac{1}{2}-k+j}\right),$$

where c_{jl} are certain constants. The hypothesis of the lemma guarantees that each summand of

$$\varphi^k(x) \sum_{j=0}^k (D^j b)(x)(1-x^2)^{-\frac{1}{2}-k+j}$$

is contained in $\mathbf{L}^2_{\varphi^{-1}}$ $(k = 0, \ldots, s)$. Hence, (7.2) is satisfied. ■

To cope with the terms K_1 and K_2 of (4.5), we now consider integral operators of the type

$$(Ku)(x) = \int_{-1}^{1} k(t,x)u(t)\,dt$$

with a kernel $k : (-1,1) \times (-1,1) \to \mathbb{C}$.

Lemma 7.27 (cf. [4], Lemma 4.2) *Let $N_s \in \mathbf{L}^1(-1,1)$, where $N_s(t) = \sigma^{-1}(t)\|k(t,\cdot)\|^2_{s,\sim}$. Then we have $K \in \mathcal{L}(\mathbf{L}^2_\sigma, \widetilde{\mathbf{L}}^2_{\sigma,s})$.*

Proof. We have

$$|\langle Ku, \widetilde{u}_n\rangle_\sigma|^2 = \left|\int_{-1}^{1}\left(\int_{-1}^{1} k(t,x)u(t)\,dt\right)\widetilde{u}_n(x)\sigma(x)\,dx\right|^2$$

$$= \left|\int_{-1}^{1}\langle k(t,\cdot), \widetilde{u}_n\rangle_\sigma u(t)\,dt\right|^2 \le \|u\|^2_\sigma \int_{-1}^{1} |\langle k(t,\cdot), \widetilde{u}_n\rangle_\sigma|^2 \sigma^{-1}(t)\,dt.$$

Thus, we can estimate

$$\|Ku\|^2_{s,\sim} = \sum_{n=0}^{\infty}(1+n)^{2s}|\langle Ku, \widetilde{u}_n\rangle_\sigma|^2$$

$$\le \sum_{n=0}^{\infty}(1+n)^{2s}\int_{-1}^{1}|\langle k(t,\cdot), \widetilde{u}_n\rangle_\sigma|^2\sigma^{-1}(t)\,dt\,\|u\|^2_\sigma$$

$$= \|u\|^2_\sigma \int_{-1}^{1}\|k(t,\cdot)\|^2_{s,\sim}\sigma^{-1}(t)\,dt,$$

and the lemma is proved. ■

Corollary 7.28 *Under the assumptions of the preceding lemma, we have $K \in \mathcal{K}(\widetilde{\mathbf{L}}^2_{\sigma,t}, \widetilde{\mathbf{L}}^2_{\sigma,s})$ for all $t > 0$, since the embedding $\widetilde{\mathbf{L}}^2_{\sigma,t} \subset \mathbf{L}^2_\sigma$ is compact.*

The operator $K_1 = bS - SbI$ is of the type we have just considered with the kernel

$$k(t,x) = -\frac{b(t)-b(x)}{t-x}.$$

Of course it is desirable to express the condition on k in Lemma 7.27 in terms of the coefficient b, but we confine ourselves here to giving a very rough sufficient condition for the boundedness of K_1 in the case $s = 1$, $\alpha, \beta > 0$..

Lemma 7.29 *Let $\alpha, \beta > 0$ and $b \in \mathbf{C}^2[-1,1]$. Then $K_1 \in \mathcal{L}(\mathbf{L}^2_\sigma, \widetilde{\mathbf{L}}^2_{\sigma,1})$.*

Proof. According to Lemma 7.27, we have to estimate $\|k(t,\cdot)\|^2_{1,\sim}$, which is equivalent to $\|w^{-1}_{\sigma^{-1}}k(t,\cdot)\|^2_{\varphi} + \|\varphi\frac{\partial}{\partial x}\left(w^{-1}_{\sigma^{-1}}k(t,\cdot)\right)\|^2_{\varphi}$ by virtue of Corollary 7.21.

The first term can be estimated by

$$\int_{-1}^{1} v^{\alpha,\beta}(x)|k(t,x)|^2\,dx \le \text{const}\,.$$

The second one is not greater than

$$\text{const}\left(\int_{-1}^{1} v^{\alpha-1,\beta-1}(x)|k(t,x)|^2\,dx + \int_{-1}^{1} v^{\alpha+1,\beta+1}(x)\left|\frac{\partial}{\partial x}k(t,x)\right|^2 dx\right).$$

At this point we remark that it can be easily shown by induction that

$$\left(\frac{\partial}{\partial x}\right)^n k(t,x) = -n!\,\frac{b(t) - \sum_{j=0}^{n}\frac{b^{(j)}(x)}{j!}(t-x)^j}{(t-x)^{n+1}}.$$

Now Taylor's formula immediately shows that the above integral is uniformly bounded in t. ■

As for $K_2 = (Sbw_\sigma^{-1} - bw_\sigma^{-1}S)w_\sigma I$, we remind that $w_\sigma I$ is an isometric isomorphism from $\mathbf{L}^2_\sigma$ onto $\mathbf{L}^2_{\varphi^{-1}}$, hence it remains to ask for conditions under which the operator $\widetilde{K_2} := S\widetilde{b}I - \widetilde{b}S$ is in $\mathcal{L}(\mathbf{L}^2_{\varphi^{-1}}, \widetilde{\mathbf{L}}^2_{\sigma,s})$, where $\widetilde{b} := w_\sigma^{-1}b$. We use the notation

$$\widetilde{k}(t,x) = \frac{\widetilde{b}(t) - \widetilde{b}(x)}{t-x}$$

for the kernel of the integral operator $\widetilde{K_2}$.

Lemma 7.30 *Let $\varphi(t)\left\|\widetilde{k}(t,\cdot)\right\|^2_{s,\sim}$ be integrable with respect to $t \in (-1,1)$. Then $\widetilde{K_2} \in \mathcal{L}(\mathbf{L}^2_{\varphi^{-1}}, \widetilde{\mathbf{L}}^2_{\sigma,s})$.*

Proof. The proof is completely analogous to that one of Lemma 7.27. ■

Of course one could easily give a condition for the boundedness of $\widetilde{K_2}$ in terms of b analogous to Lemma 7.29 (cf. also Lemma 4.21).

7.4 Mapping properties of the operator bS in the special case $v = \varphi$, $\sigma = \varphi^{-1}$

When we consider the boundedness of the operator bS, $b \in \mathbf{PC}$, in the spaces $\widetilde{\mathbf{L}}^2_{\sigma,s}$, the special case $v = \varphi$, $\sigma = \varphi^{-1}$ allows again a considerable simplification, since we need not make use of the decomposition (4.5). The relation

$$S\widetilde{u}_n = S\varphi U_n = iT_{n+1}$$

shows that $S \in \mathcal{L}(\widetilde{\mathbf{L}}^2_{\sigma,s}, \mathbf{L}^2_{\varphi^{-1},s})$ for all $s \ge 0$. Thus it remains to find conditions on b such that $bI \in \mathcal{L}(\mathbf{L}^2_{\varphi^{-1},s}, \widetilde{\mathbf{L}}^2_{\sigma,s})$. For integer s, this is done by the following lemma.

Lemma 7.31 ([33], Lemma 8.1) *Let $s \geq 0$ be an integer and assume that $\varphi^{k+1}D^k(\varphi^{-1}b) \in \mathbf{L}^\infty$ for all $k = 0, \dots, s$. Then we have $bI \in \mathcal{L}(\mathbf{L}^2_{\varphi^{-1},s}, \widetilde{\mathbf{L}}^2_{\sigma,s})$.*

Proof. First recall that $\varphi^{-1}I$ is an isometric isomorphism from $\widetilde{\mathbf{L}}^2_{\sigma,s}$ onto $\mathbf{L}^2_{\varphi,s}$. Furthermore, we have in view of Theorem 7.20

$$\|bu\|_{s,\sim} = \|b\varphi^{-1}u\|_{\varphi,s} \sim \sum_{k=0}^{s} \|\varphi^k D^k(b\varphi^{-1}u)\|_\varphi = \sum_{k=0}^{s} \|\varphi^{k+1}D^k(b\sigma u)\|_{\varphi^{-1}}$$

$$\leq \sum_{k=0}^{s}\sum_{j=0}^{k} \binom{k}{j} \|\varphi^{j+1}D^j(\varphi^{-1}b)\|_\infty \|\varphi^{k-j}D^{k-j}u\|_\sigma$$

$$\leq \text{const} \sum_{k=0}^{s}\sum_{j=0}^{k} \binom{k}{j} \|\varphi^{j+1}D^j(\varphi^{-1}b)\|_\infty \|u\|_{\sigma,s} \quad \blacksquare$$

7.5 Invertibility of singular integral operators in $\widetilde{\mathbf{L}}^2_{\sigma,s}$

We conclude the present section by giving some results on the invertibility of singular integral operators in Sobolev spaces. For this end, we have to provide some material from [20].

In what follows, let $\Gamma \subset \mathbb{C}$ be a simple closed Lyapunov curve. Then Γ divides the complex plane into an inner domain F_Γ^+ and an outer domain F_Γ^-. Denote by $\mathbf{C}_+(\Gamma)$ the set of all functions being holomorphic in F_Γ^+ and continuous on $\overline{F_\Gamma^+}$. The set $\mathbf{C}_-(\Gamma)$ consists of all functions holomorphic in F_Γ^- and continuous on $\overline{F_\Gamma^-}$ that additionally vanish at infinity.

Definition 7.32 *We say that a function $c \in \mathcal{G}\mathbf{C}(\Gamma)$ allows a factorization if there exist functions $c_+ \in \mathbf{C}_+(\Gamma)$, $c_- \in \mathbf{C}_-(\Gamma)$, points $t_+ \in F_\Gamma^+$, $t_- \in F_\Gamma^-$ and an integer κ such that*

$$c(t) = c_+(t)\left(\frac{t-t_+}{t-t_-}\right)^\kappa c_-(t) \tag{7.3}$$

for all $t \in \Gamma$, where $c_\pm(t) \neq 0$ for all $t \in \Gamma$. The number κ is called index of c.

Lemma 7.33 (cf. [20], Th. III.3.1 and preceding examples) *Assume that the curve Γ is infinitely smooth, and let $c \in \mathbf{C}^{m,\eta}(\Gamma)$, $m \geq 0$, $0 < \eta < 1$. If $c(t) \neq 0$ for all $t \in \Gamma$, then c allows a factorization of the form (7.3), where the factors c_+, c_- are again in $\mathbf{C}^{m,\eta}(\Gamma)$.*

We introduce a weight $\varrho(t) = \prod_{j=1}^{m} |t-t_j|^{\gamma_j}$, where $t_1, \dots, t_m \in \Gamma$, $-1 < \gamma_j < 1$. This ensures $S_\Gamma \in \mathcal{L}(\mathbf{L}^2_\varrho(\Gamma))$. Note that a factorization of the form (7.3) is at the same time a generalized $\mathbf{L}^2_\varrho(\Gamma)$-factorization in the sense of [20, VIII, §1].

Proposition 7.34 (cf. [20], Th. IX.3.1, Folgerung IX.3.1) *Let $f, g \in \mathbf{C}(\Gamma)$. The operator $fP_\Gamma + gQ_\Gamma$ is invertible in $\mathbf{L}^2_\varrho(\Gamma)$ if and only if $f, g \in \mathcal{G}\mathbf{C}(\Gamma)$ and $g^{-1}f$ allows a generalized $\mathbf{L}^2_\varrho(\Gamma)$-factorization in the sense of [20, VIII, §1] with the index $\kappa = 0$.*

Now we consider the case of a singular integral operator $fP_\Gamma + gQ_\Gamma$ on an open Lyapunov curve Γ. In what follows we assume that f and g are continuous on Γ and

$$f(t_1) = g(t_1) \neq 0, \quad f(t_2) = g(t_2) \neq 0, \tag{7.4}$$

where t_1 and t_2 are the endpoints of Γ. We extend Γ to a smooth closed Lyapunov curve $\widetilde{\Gamma}$ and define the coefficients $\widetilde{f}, \widetilde{g}$ and the weight $\widetilde{\varrho}$ on $\widetilde{\Gamma}$ by

$$\widetilde{f}(t) = \begin{cases} f(t), & t \in \Gamma \\ 1, & t \in \widetilde{\Gamma} \setminus \Gamma, \end{cases}$$

and analogously for $\widetilde{g}$ and $\widetilde{\varrho}$.

Remark 7.35 *Condition (7.4) ensures that $\widetilde{g}^{-1}\widetilde{f}$ is continuous on $\widetilde{\Gamma}$, which allows to apply the preceding results concerning the case of continuous coefficients together with Lemma 7.36 to the operator $g(g^{-1}fP_\Gamma + Q_\Gamma)$.*

Lemma 7.36 ([20], Folgerung VII.1.3) *The operator $fP_\Gamma + gQ_\Gamma$ is invertible in $\mathbf{L}^2_\varrho(\Gamma)$ if and only if $\widetilde{f}P_{\widetilde{\Gamma}} + \widetilde{g}Q_{\widetilde{\Gamma}}$ is invertible in $\mathbf{L}^2_{\widetilde{\varrho}}(\widetilde{\Gamma})$.*

Now assume that $fP_\Gamma + gQ_\Gamma$ is invertible on $\mathbf{L}^2_\varrho(\Gamma)$ and $\widetilde{c} = \widetilde{c}_+\widetilde{c}_-$ is the factorization of $\widetilde{c} = \widetilde{g}^{-1}\widetilde{f}$ that exists by virtue of Proposition 7.34.

Lemma 7.37 ([20], Theorem IX.4.2) *Let $fP_\Gamma + gQ_\Gamma$ be invertible in $\mathbf{L}^2_\varrho(\Gamma)$, where Γ is an open Lyapunov curve. Then the inverse operator is given by*

$$A^{-1} = c_- f^{-1}(gP_\Gamma + fQ_\Gamma)g^{-1}c_-^{-1}I,$$

where c_- is the restriction of $\widetilde{c}_-$ onto Γ.

Theorem 7.38 *Let $s \geq 1$ be an integer and a, b functions such that the operators aI and bS are bounded in $\widetilde{\mathbf{L}}^2_{\sigma,s}$. Moreover, assume that $a, b \in \mathbf{C}^{s,\eta}[-1,1]$, $0 < \eta < 1$, and*

$$(D^j b)(\pm 1) = 0, \qquad j = 0, \ldots, s.$$

Then the invertibility of $A = aI + bS$ in $\mathbf{L}^2_\sigma$ implies that $A^{-1} \in \mathcal{L}(\widetilde{\mathbf{L}}^2_{\sigma,s})$.

Proof. Since $A = (a+b)P + (a-b)Q$ is invertible, we have $(a+b)(t) \neq 0$ and $(a-b)(t) \neq 0$ for all $t \in [-1,1]$. Put $c := \frac{a+b}{a-b}$. Then $c \in \mathbf{C}^{s,\eta}[-1,1]$.

First we verify that $(D^k c)(\pm 1) = 0$ for $k = 1, \ldots s$. If we abbreviate $f := a+b$, $g := a-b$, the k-th derivative of c $(k = 1, \ldots, s)$ is a fraction the denominator of which equals g^{2k}, whereas the enumerator is a sum of products each of which contains a derivative of $f'g - fg'$. Evidently, for $m = 0, \ldots, s-1$ the expression

$$D^m(f'g - fg') = \sum_{j=0}^{m} \binom{m}{j} \left(D^{j+1}fD^{m-j}g - D^{j+1}gD^{m-j}f\right)$$

vanishes in ± 1, since we have $(D^j f)(\pm 1) = (D^j g)(\pm 1) = (D^j a)(\pm 1)$, $j = 0, \ldots, s$. Hence, $(D^k c)(\pm 1) = 0$ for $k = 1, \ldots, s$.

Extend $[-1, 1]$ to an infinitely smooth closed curve $\widetilde{\Gamma}$ and put

$$\widetilde{c}(t) = \begin{cases} c(t), & t \in [-1, 1] \\ 1, & t \in \widetilde{\Gamma} \setminus [-1, 1]. \end{cases}$$

Then the previous consideration shows that $\widetilde{c} \in \mathbf{C}^{s,\eta}(\widetilde{\Gamma})$. Lemma 7.36 and Proposition 7.34 imply that $\widetilde{c}$ allows a factorization, where the factors $\widetilde{c}_+$, $\widetilde{c}_-$ belong to $\mathbf{C}^{s,\eta}(\widetilde{\Gamma})$ due to Lemma 7.33. Hence, we also have $c_- = \widetilde{c}_-|_{[-1,1]} \in \mathbf{C}^{s,\eta}[-1, 1]$. Lemma 7.37 states that

$$\begin{aligned} A^{-1} &= c_-(a+b)^{-1}((a-b)P + (a+b)Q)(a-b)^{-1}c_-^{-1}I \\ &= c_-(a+b)^{-1}(aI - bS)(a-b)^{-1}c_-^{-1}I, \end{aligned}$$

which is clearly an element of $\mathcal{L}(\widetilde{\mathbf{L}}^2_{\sigma,s})$ (compare Corollary 7.25). ■

8 Implementation of the collocation method and numerical results

8.1 Fast solution of the linear approximate equations in special cases

For certain combinations of the weights v and σ, a suitable implementation of the collocation method enables us to solve the resulting system of linear equations with a fast algorithm that requires only $O(n^2)$ operations and $O(n)$ storage due to the special structure of the system matrix.

We restrict ourselves to considering the case $\sigma = v^{-1}$, that is, we have $w_{v,\sigma^{-1}} = v$. We search for the values of the approximate solution u_n of (2.5) in the collocation points x^v_{kn}, $k = 1, \ldots, n$, that is, the zeros of p^v_n. We now write the weighted polynomial u_n in the form

$$u_n(x) = w_{v,\sigma^{-1}}(x)w_n(x) = v(x)\sum_{k=1}^{n} \xi_{kn} l^v_{kn}(x), \tag{8.1}$$

where $\xi_{kn} = w_n(x^v_{kn})$ and

$$l^v_{kn}(x) = \frac{\widetilde{p}^v_n(x)}{(x - x^v_{kn})(\widetilde{p}^v_n)'(x^v_{kn})}$$

is the k-th fundamental polynomial of Lagrange interpolation with respect to the nodes x^v_{kn}, and $\widetilde{p}^v_n$ denotes the monic orthogonal polynomial of degree n with respect to the weight v. (Of course we could also have taken the normalized orthogonal polynomials here, but we choose the monic ones because the recurrence relation (8.3) looks somewhat easier for them.)

Let us consider now the action of the operator S on a weighted polynomial of the form (8.1). Let first $x \neq x^v_{kn}$ for all $k = 1, \ldots, n$. Then

$$(Svw_n)(x) = \sum_{k=1}^{n} \xi_{kn} \frac{1}{\pi i}\int_{-1}^{1} \frac{v(t)\widetilde{p}^v_n(t)}{(t - x^v_{kn})(\widetilde{p}^v_n)'(x^v_{kn})(t-x)}\,\mathrm{d}t$$

$$= \sum_{k=1}^{n} \frac{\xi_{kn}}{(\widetilde{p}^v_n)'(x^v_{kn})(x^v_{kn} - x)} \frac{1}{\pi i}\int_{-1}^{1} v(t)\widetilde{p}^v_n(t)\left(\frac{1}{t - x^v_{kn}} - \frac{1}{t-x}\right)\mathrm{d}t$$

$$= \sum_{k=1}^{n} \frac{\varrho^v_n(x^v_{kn}) - \varrho^v_n(x)}{x^v_{kn} - x} \frac{1}{(\widetilde{p}^v_n)'(x^v_{kn})}\xi_{kn},$$

where $\varrho^v_n(x) = (Sv\widetilde{p}^v_n)(x)$. Now we can write down equation (2.5) in the form

$$a(x^v_{jn})v(x^v_{jn})w_n(x^v_{jn}) + b(x^v_{jn})(Svw_n)(x^v_{jn})$$

$$= \sum_{k=1}^{n} a_{jk}\frac{1}{(\widetilde{p}^v_n)'(x^v_{kn})}\xi_{kn} = f(x^v_{jn}), \qquad j = 1, \ldots, n,$$

where

$$a_{jk} = \begin{cases} \frac{a_k b_j - a_j b_j}{c_k - c_j}, & k \neq j \\ d_j, & k = j \end{cases} \qquad j,k = 1,\ldots,n, \tag{8.2}$$

with

$$a_k = \varrho_n^v(x_{kn}^v),$$

$$b_j = b(x_{jn}^v),$$

$$c_k = x_{kn}^v,$$

$$d_j = a(x_{jn}^v)v(x_{jn}^v)(\widetilde{p}_n^v)'(x_{jn}^v) + b(x_{jn}^v)(\varrho_n^v)'(x_{jn}^v).$$

For the computation of $(\widetilde{p}_n^v)'$ we make use of the relation

$$\left(\widetilde{p}_n^{v^{\gamma,\delta}}\right)' = n\,\widetilde{p}_{n-1}^{v^{\gamma+1,\delta+1}}$$

(cf. [51, Equ. 4.21.7]). We remark that in the cases where $|\gamma| = |\delta| = \frac{1}{2}$ the expressions for ϱ_n^v and $(\varrho_n^v)'$ as well as $(\widetilde{p}_n^v)'$ are explicitly known (cf. [44, 9.15]). In the other cases, they can be determined as indicated in the sequel.

It is well-known that the monic orthogonal polynomials satisfy a three-term recurrence formula of the form

$$\widetilde{p}_{k+1}^v(x) = (x - \alpha_k)\widetilde{p}_k^v(x) - \beta_k \widetilde{p}_{k-1}^v(x), \qquad k = 0,1,2,\ldots, \tag{8.3}$$

where $\widetilde{p}_{-1}^v \equiv 0$ and $\widetilde{p}_0^v \equiv 1$. In the case of Jacobi weights $v = v^{\gamma,\delta}$, there are explicit formulas for the α_k and β_k:

$$\alpha_0 = \frac{\delta - \gamma}{\gamma + \delta + 2}, \qquad \beta_1 = \frac{4(\gamma+1)(\delta+1)}{(\gamma+\delta+2)^2(\gamma+\delta+3)},$$

$$\alpha_k = \frac{\delta^2 - \gamma^2}{(\gamma+\delta+2k)(\gamma+\delta+2k+2)}, \qquad k = 1,2,\ldots,$$

$$\beta_k = \frac{4(\gamma+k)(\delta+k)(\gamma+\delta+k)k}{(\gamma+\delta+2k)^2(\gamma+\delta+2k+1)(\gamma+\delta+2k-1)}, \qquad k = 2,3,\ldots.$$

This allows us to compute the ϱ_n^v, $(\varrho_n^v)'$ recursively:

Lemma 8.1 (cf. [18]) *Let $\varrho_k^v = Sv\widetilde{p}_k^v$. Then we have the recurrence relations*

$$\varrho_{k+1}^v(x) = (x - \alpha_k)\varrho_k^v(x) - \beta_k \varrho_{k-1}^v(x), \qquad k = 1,\ldots,n-1,$$

and

$$(\varrho_{k+1}^v)'(x) = (x - \alpha_k)(\varrho_k^v)'(x) + \varrho_k^v(x) - \beta_k(\varrho_{k-1}^v)'(x), \qquad k = 1,\ldots,n-1,$$

with the initial values

$$\varrho_1^v(x) = \beta_0 + (x - \alpha_0)\varrho_0^v(x)$$

and

$$(\varrho_1^v)'(x) = \varrho_0^v(x) + (x - \alpha_0)(\varrho_0^v)'(x),$$

where

$$\beta_0 := \frac{1}{\pi i}\int_{-1}^{1} v(t)\,\mathrm{d}t.$$

(For the computation of ϱ_0^v and $(\varrho_0^v)'$ see below.)

Proof. It is sufficient to show a recurrence relation for ϱ_k^v, from which the corresponding relations for $(\varrho_k^v)'$ follow immediately. We have

$$\varrho_1^v(x) = \frac{1}{\pi i}\int_{-1}^{1} \frac{t-\alpha_0}{t-x} v(t)\,\mathrm{d}t = \beta_0 + (x-\alpha_0)\varrho_0^v(x),$$

and for $k \geq 1$, (8.3) yields

$$\varrho_{k+1}^v(x) = \frac{1}{\pi i}\int_{-1}^{1} \frac{(t-\alpha_k)\widetilde{p}_k^v(t)}{t-x} v(t)\,\mathrm{d}t - \frac{\beta_k}{\pi i}\int_{-1}^{1} \frac{\widetilde{p}_{k-1}^v(t)}{t-x} v(t)\,\mathrm{d}t$$

$$= \frac{1}{\pi i}\int_{-1}^{1} \widetilde{p}_k^v(t) v(t)\,\mathrm{d}t + (x-\alpha_k)\varrho_k^v(x) - \beta_k \varrho_{k-1}^v(x),$$

where the first term vanishes for $k \geq 1$ because of the orthogonality relations. ■

We remark that

$$\beta_0 = \frac{2^{\gamma+\delta+1}}{\pi i}\frac{\Gamma(\gamma+1)\,\Gamma(\delta+1)}{\Gamma(\gamma+\delta+2)}.$$

The computation of ϱ_0^v is based upon the following relation ([18, Equ. (2.2)]) that holds for $\gamma \neq 0$ and $x \in (-1,1)$:

$$\begin{aligned}\pi i \varrho_0^v(x) &= (1-x)^\gamma(1+x)^\delta \pi \cot\gamma\pi - 2^{\gamma+\delta}\tfrac{\Gamma(\gamma)\Gamma(\delta+1)}{\Gamma(\gamma+\delta+1)}\, {}_2F_1(1, -\gamma-\delta; 1-\gamma; \tfrac{1-x}{2}) \\ &= (1+x)^\delta\left[(1-x)^\gamma \pi\cot\gamma\pi + 2^\gamma\tfrac{\Gamma(\gamma+1)\Gamma(\delta+1)}{\Gamma(\gamma+\delta+1)}\sum_{k=0}^{\infty}\tfrac{(1+\delta)_k}{(k-\gamma)k!}\left(\tfrac{1-x}{2}\right)^k\right],\end{aligned}$$

where

$${}_2F_1(A,B;C;z) := \sum_{k=0}^{\infty}\frac{(A)_k(B)_k}{(C)_k}\frac{z^k}{k!} \qquad ((A)_0 := 1,\ (A)_{k+1} := (A+k)(A)_k)$$

denotes the hypergeometric series, which converges for $|z| < 1$.

Hence, we can compute the derivative of ϱ_0^v by

$$\pi i(\varrho_0^v)'(x) = \pi\cot\gamma\pi\left[-\gamma(1-x)^{\gamma-1}(1+x)^\delta + \delta(1-x)^\gamma(1+x)^{\delta-1}\right]$$

$$+2^{\gamma+\delta-1}\frac{\Gamma(\gamma)\Gamma(\delta+1)}{\Gamma(\gamma+\delta+1)}\sum_{k=1}^{\infty}\frac{(-\gamma-\delta)_k}{(1-\gamma)_k}k\left(\frac{1-x}{2}\right)^{k-1}.$$

Since the convergence of the hypergeometric series is too slow for z near 1 (which means x near -1), we use the relations

$$\varrho_0^{v^{\gamma,\delta}}(x) = -\varrho_0^{v^{\delta,\gamma}}(-x), \qquad (\varrho_0^{v^{\gamma,\delta}})'(x) = (\varrho_0^{v^{\delta,\gamma}})'(-x)$$

for negative x if also $\delta \neq 0$. (Note that if $\gamma+\delta = -1$, the expression $\Gamma(\gamma+\delta+1)$ becomes infinite, and the second summand in brackets vanishes.) The case $\gamma \neq 0$, $\delta = 0$ (and vice versa) is excluded from our considerations. In the special case $\gamma = \delta = 0$, that is, $v \equiv 1$, we finally have $\varrho_0^v(x) = \frac{1}{\pi i} \ln \frac{1-x}{1+x}$.

In the following we will show how to solve efficiently a system of linear equations with a Löwner-like matrix $\mathbf{A}_n$ the entries of which are of the form (8.2). Having solved this system, we only have to multiply its solution by the diagonal matrix $\operatorname{diag}\left((\widetilde{p}_n^v)'(x_{kn}^v)\right)_{k=1}^n$ to obtain the ξ_{kn}.

We assume that $\mathbf{A}_n$ is strongly regular, that means all sections $\mathbf{A}_m = (a_{jk})_{j,k=1}^m$ are invertible for $m = 1, \ldots, n$. The fast algorithm is essentially based on the following two lemmas (compare [23]), the first of which is valid for arbitrary matrices.

Lemma 8.2 *Let $\mathbf{A}_n = (a_{jk})_{j,k=1}^n$ be regular, and let the vectors $x^{n-1} = (x_k^{n-1})_{k=1}^{n-1}$ and $x_0^{n-1} = (x_{0k}^{n-1})_{k=1}^{n-1}$ be solutions of*

$$\mathbf{A}_{n-1}x^{n-1} = (f_j)_{j=1}^{n-1} \text{ and } \mathbf{A}_{n-1}x_0^{n-1} = (a_{jn})_{j=1}^{n-1}.$$

Then the solution x^n of $\mathbf{A}_n x^n = (f_j)_{j=1}^n$ is given by

$$x^n = \begin{pmatrix} x^{n-1} \\ 0 \end{pmatrix} + \frac{\gamma_n}{\eta_n} \begin{pmatrix} x_0^{n-1} \\ -1 \end{pmatrix},$$

where

$$\gamma_n = f_n - \sum_{k=1}^{n-1} a_{nk} x_k^{n-1}, \qquad \eta_n = \sum_{k=1}^{n-1} a_{nk} x_{0k}^{n-1} - a_{nn}.$$

(Note that the regularity of $\mathbf{A}_n$ implies $\eta_n \neq 0$.) The special structure of the matrix with the entries (8.2) allows us to determine the vectors x_0^m recursively from the solutions x_1^m, x_2^m of the fundamental equations

$$\mathbf{A}_m x_1^m = (b_j)_{j=1}^m \text{ and } \mathbf{A}_m x_2^m = (-a_j b_j)_{j=1}^m.$$

Lemma 8.3 (cf. [7], Equ. (2.31)) *For $m = 1, \ldots, n-1$ let*

$$\beta_{m1} = 1 - \sum_{k=1}^m \frac{a_{m+1} - a_k}{c_{m+1} - c_k} x_{1k}^m, \qquad \beta_{m2} = a_{m+1} + \sum_{k=1}^m \frac{a_{m+1} - a_k}{c_{m+1} - c_k} x_{2k}^m,$$

and

$$\alpha_m = \beta_{m1} \left[1 + \sum_{k=1}^m \frac{x_{2k}^m}{c_{m+1} - c_k} \right] + \beta_{m2} \sum_{k=1}^m \frac{x_{1k}^m}{c_{m+1} - c_k}.$$

Then

$$x_{0k}^m = \frac{\beta_{m2} x_{1k}^m + \beta_{m1} x_{2k}^m}{\alpha_m (c_{m+1} - c_k)}, \qquad k = 1, \ldots, m.$$

Proof. For $j = 1, \ldots, m$ we have

$$\sum_{k=1}^{m} a_{jk} x_{0k}^m = \frac{\beta_{m2}}{\alpha_m} \left[d_j \frac{x_{1j}^m}{c_{m+1} - c_j} + \sum_{k \neq j} \frac{a_k b_j - a_j b_j}{c_k - c_j} \frac{x_{1k}^m}{c_{m+1} - c_k} \right]$$

$$+ \frac{\beta_{m1}}{\alpha_m} \left[d_j \frac{x_{2j}^m}{c_{m+1} - c_j} + \sum_{k \neq j} \frac{a_k b_j - a_j b_j}{c_k - c_j} \frac{x_{2k}^m}{c_{m+1} - c_k} \right]. \tag{8.4}$$

The first summand of this expression equals

$$\frac{\beta_{m2}}{\alpha_m (c_{m+1} - c_j)} \left[d_j x_{1j}^m + \sum_{k \neq j} \left(\frac{1}{c_k - c_j} + \frac{1}{c_{m+1} - c_k} \right) (a_k b_j - a_j b_j) x_{1k}^m \right]$$

$$= \frac{\beta_{m2}}{\alpha_m (c_{m+1} - c_j)} \left[d_j x_{1j}^m + b_j - d_j x_{1j}^m + \sum_{k \neq j} \frac{a_k b_j - a_j b_j}{c_{m+1} - c_k} x_{1k}^m \right]$$

$$= \frac{\beta_{m2}}{\alpha_m (c_{m+1} - c_j)} \left[b_j + (a_{m+1} b_j - a_j b_j) \sum_{k \neq j} \frac{x_{1k}^m}{c_{m+1} - c_k} - \sum_{k \neq j} \frac{a_{m+1} b_j - a_k b_j}{c_{m+1} - c_k} x_{1k}^m \right]$$

$$= \frac{\beta_{m2}}{\alpha_m (c_{m+1} - c_j)} \left[b_j + (a_{m+1} b_j - a_j b_j) \sum_{k=1}^{m} \frac{x_{1k}^m}{c_{m+1} - c_k} - \sum_{k=1}^{m} \frac{a_{m+1} b_j - a_k b_j}{c_{m+1} - c_k} x_{1k}^m \right]$$

$$= \frac{\beta_{m2}}{\alpha_m (c_{m+1} - c_j)} \left[b_j \beta_{m1} + (a_{m+1} b_j - a_j b_j) \sum_{k=1}^{m} \frac{x_{1k}^m}{c_{m+1} - c_k} \right].$$

In the same manner one can verify that the second summand equals

$$\frac{\beta_{m1}}{\alpha_m (c_{m+1} - c_j)} \left[-a_j b_j + (a_{m+1} b_j - a_j b_j) \sum_{k=1}^{m} \frac{x_{2k}^m}{c_{m+1} - c_k} - \sum_{k=1}^{m} \frac{a_{m+1} b_j - a_k b_j}{c_{m+1} - c_k} x_{2k}^m \right]$$

$$= \frac{\beta_{m1}}{\alpha_m (c_{m+1} - c_j)} \left[-a_j b_j + (a_{m+1} b_j - a_j b_j) \left(1 + \sum_{k=1}^{m} \frac{x_{2k}^m}{c_{m+1} - c_k} \right) - a_{m+1} b_j + a_j b_j \right.$$

$$\left. - \sum_{k=1}^{m} \frac{a_{m+1} b_j - a_k b_j}{c_{m+1} - c_k} x_{2k}^m \right]$$

$$= \frac{\beta_{m1}}{\alpha_m (c_{m+1} - c_j)} \left[(a_{m+1} b_j - a_j b_j) \left(1 + \sum_{k=1}^{m} \frac{x_{2k}^m}{c_{m+1} - c_k} \right) - b_j \beta_{m2} \right].$$

Consequently, (8.4) is equal to

$$\frac{a_{m+1} b_j - a_j b_j}{c_{m+1} - c_j} \left[\frac{\beta_{m2}}{\alpha_m} \sum_{k=1}^{m} \frac{x_{1k}^m}{c_{m+1} - c_k} + \frac{\beta_{m1}}{\alpha_m} \left(1 + \sum_{k=1}^{m} \frac{x_{2k}^m}{c_{m+1} - c_k} \right) \right]$$

$$= \frac{a_{m+1} b_j - a_j b_j}{c_{m+1} - c_j} \qquad \blacksquare$$

Now we can alternately apply Lemma 8.2 and Lemma 8.3 to obtain the following algorithm:

Algorithm 8.4

- Put $x_1^1 := f_1/a_{11}$, $x_{01}^1 := a_{12}/a_{11}$, $x_{11}^1 := b_1/a_{11}$, $x_{21}^1 := -a_1 b_1/a_{11}$
- FOR $m := 2$ TO $n-1$ DO
  - Compute x^m, x_1^m, x_2^m from x^{m-1}, x_1^{m-1}, x_2^{m-1}, and x_0^{m-1} by Lemma 8.2
  - Compute x_0^m from x_1^m and x_2^m by Lemma 8.3
- Compute x^n by Lemma 8.2

To evaluate the results of our numerical tests, we compute the error

$$\varepsilon_{n,t} = \|u_n - P_n^\sigma u\|_{t,\sim} = \left(\sum_{k=0}^{n-1} (1+k)^{2t} |c_{kn} - c_k|^2 \right)^{\frac{1}{2}}$$

(see Section 7, in particular $\varepsilon_{n,0} = \|u_n - P_n^\sigma u\|_\sigma$), where

$$c_{kn} = \langle u_n, \widetilde{u}_k \rangle_\sigma, \qquad c_k = \langle u, \widetilde{u}_k \rangle_\sigma$$

are the Fourier coefficients of the approximate and the exact solution, respectively. It is easy to compute the c_{kn}, $k = 0, \ldots, n-1$, from the function values ξ_{jn}, $j = 1, \ldots, n$ using the Gaussian quadrature formula. We have

$$c_{kn} = \int_{-1}^{1} w_n(x) v(x) p_k^v(x)\, \mathrm{d}x = \sum_{j=1}^{n} A_{jn}^v \, \xi_{jn} \, p_k^v(x_{jn}^v).$$

If we note that for the normalized orthogonal polynomials the recurrence relation

$$\sqrt{\beta_{k+1}}\, p_{k+1}^v(x) = (x - \alpha_k) p_k^v(x) - \sqrt{\beta_k}\, p_{k-1}^v(x), \qquad k = 0, 1, \ldots$$

with $p_{-1} \equiv 0$, $p_0 \equiv \frac{1}{\sqrt{\beta_0}}$ holds, we obtain the following algorithm for the determination of the c_{kn}:

Algorithm 8.5

- Put $b_{j0} = A_{jn}^v \, \xi_{jn} \frac{1}{\sqrt{\beta_0}}$, $\quad b_{j1} = A_{jn}^v \, \xi_{jn} \frac{1}{\sqrt{\beta_0 \beta_1}} (x_{jn}^v - \alpha_0)$, $\quad j = 1, \ldots, n$
- Compute $c_{0n} = \sum_{j=1}^n b_{j0}$, $\qquad c_{1n} = \sum_{j=1}^n b_{j1}$
- FOR $k := 1$ TO $n-2$ DO
  - Put $b_{j,k+1} = \frac{1}{\sqrt{\beta_{k+1}}} [(x_{jn}^v - \alpha_k) b_{jk} - \sqrt{\beta_k}\, b_{j,k-1}]$, $\quad j = 1, \ldots, n$
  - Compute $c_{k+1,n} = \sum_{j=1}^n b_{j,k+1}$

For the determination of the weights A_{kn}^v as well as of the nodes x_{kn}^v in the case of arbitrary Jacobi weights v we refer to [44, 9.39]. In the special cases $|\gamma| = |\delta| = \frac{1}{2}$, to which we restricted ourselves in the numerical tests, there are explicit expressions for these parameters.

8.2 Numerical examples

Let us now give some numerical examples. We only consider the case $v = \varphi$, $\sigma = \varphi^{-1}$.

$$(\mathbf{A}) \qquad a(x) = \operatorname{sgn} x, \qquad b(x) = ix, \qquad u(x) = |x|$$

$$(\mathbf{B}) \qquad a(x) = 2, \qquad b(x) = i(1 - x^2), \qquad u(x) = 1 - x^2$$

We remark that $u \in \widetilde{\mathbf{L}}^2_{\sigma,s}$ for $s < \frac{1}{2}$ in Example (A) and for $s < \frac{5}{2}$ in Example (B).

Example (A)		
n	$\varepsilon_{n,0}$	$n^{1/2}\,\varepsilon_{n,0}$
20	0.1444968667	0.646210
40	0.1050036958	0.664102
60	0.0865353255	0.670300
80	0.0752951173	0.673460
100	0.0675380083	0.675380
500	0.0304858965	0.681685
1000	0.0215824621	0.682497
2000	0.0152702514	0.682906
4000	0.0108009463	0.683112

Example (B)		
n	$\varepsilon_{n,0}$	$n^{5/2}\,\varepsilon_{n,0}$
20	0.0004078075	0.729508
40	0.0000813336	0.823038
60	0.0000307953	0.858743
80	0.0000153296	0.877519
100	0.0000088909	0.889090
500	0.0000001660	0.927706
1000	0.0000000295	0.932707
2000	0.0000000052	0.935222
4000	0.0000000009	0.936483

These examples show the influence of the smoothness of the solution on the convergence rate, even in cases where the author could not predict the corresponding rate theoretically (in particular, in Example (A), the hypotheses of Theorem 7.19 are not satisfied).

$$(\mathbf{C}) \qquad a(x) = \sqrt{1 - x^2}, \qquad b(x) = -ix, \qquad u(x) = |x|\sqrt{1 - x^2}$$

In Example (C) we have $u \in \widetilde{\mathbf{L}}^2_{\sigma,s}$ for $s < \frac{3}{2}$.

Example (C)		
n	$\varepsilon_{n,0}$	$n^{3/2}\,\varepsilon_{n,0}$
20	0.0057887340	0.517760
40	0.0023048099	0.583076
60	0.0013106439	0.609132
80	0.0008708601	0.623137
100	0.0006318786	0.631879
500	0.0000591851	0.661709
1000	0.0000210497	0.665650
2000	0.0000074644	0.667639
4000	0.0000026430	0.668639

In Example (C), the operator $\widetilde{A} = aI - bS$ is not invertible. Nevertheless, we can observe convergence. This can be explained if we remember that the collocation equations (2.5) can be interpreted in a different way if we apply L_n^φ instead of $\widetilde{L_n^\sigma}$ in (2.6) (cf. Remark 2.1). Further, we note that the operator $A = \varphi I - ieS$ in Example (C) has the property

$$Au = \sqrt{2}\langle u, \widetilde{u}_0\rangle_\sigma T_0 + \sum_{n=1}^{\infty}\langle u, \widetilde{u}_n\rangle_\sigma T_n,$$

that is, A maps X_n onto Π_n. Thus, in this particular case, Equation (2.6) is equivalent to

$$u_n = A^{-1}L_n^\varphi f,$$

which converges to the solution of the original equation for all right-hand sides f that satisfy $\|f - L_n^\varphi f\|_\sigma \to 0 \quad (n \to \infty)$.

8.3 Solution of the approximate equations in the general case

If we do not have the special case $\sigma = v^{-1}$ considered in the preceding subsection, we proceed in a different way. We write the weighted polynomial u_n that solves (2.5) in the form

$$u_n(x) = w_{v,\sigma^{-1}}(x)\sum_{k=0}^{n-1}\xi_{kn}\widetilde{p}_k^w(x),$$

where $\widetilde{p}_k^w$ denotes the *monic* orthogonal polynomial of degree k with respect to the weight $w := w_{v,\sigma^{-1}}$, and we solve the system of linear equations

$$\sum_{k=0}^{n-1}\left[\left(aw_{v,\sigma^{-1}}\widetilde{p}_k^w\right)(x_{jn}^v) + b(x_{jn}^v)\left(Sw_{v,\sigma^{-1}}\widetilde{p}_k^w\right)(x_{jn}^v)\right]\xi_{kn}$$

$$=: \sum_{k=0}^{n-1} a_{jk}\,\xi_{kn} = f(x_{jn}^v), \qquad j = 1,\dots,n \tag{8.5}$$

to determine the ξ_{kn}, $k = 0,\dots,n-1$.

The recurrence formula (8.3) allows us to compute the matrix coefficients a_{jk} recursively in an analogous way as in Lemma 8.1, where the recurrence coefficients α_k, β_k in the following lemma correspond to the weight $w_{v,\sigma^{-1}}$:

Lemma 8.6 *We have the recurrence relation*

$$a_{j,k+1} = (x_{jn}^v - \alpha_k)a_{jk} - \beta_k a_{j,k-1}, \qquad k = 1,\dots,n-1,$$

with the initial values

$$a_{j0} = \left(aw_{v,\sigma^{-1}}\right)(x_{jn}^v) + b(x_{jn}^v)\varrho_0^w(x_{jn}^v)$$

and

$$\begin{aligned} a_{j1} &= (x^v_{jn} - \alpha_0)\,(aw_{v,\sigma^{-1}})\,(x^v_{jn}) + b(x^v_{jn})\,\big(\beta_0 + (x^v_{jn} - \alpha_0)\varrho^w_0(x^v_{jn})\big) \\ &= (x^v_{jn} - \alpha_0)a_{j0} + b(x^v_{jn})\beta_0, \end{aligned}$$

where

$$\beta_0 := \frac{1}{\pi i}\int_{-1}^{1} w_{v,\sigma^{-1}}(t)\,\mathrm{d}t.$$

If (8.5) has been solved, one can efficiently compute function values of u_n from the coefficients ξ_{kn} using (8.3) in the following way:

Algorithm 8.7

- Put $b_0 := 1$, $b_1 := x - \alpha_0$, $s := \xi_{0n}b_0 + \xi_{1n}b_1$
- FOR $k := 1$ TO $n-2$ DO
  - Put $b_{k+1} := (x - \alpha_k)b_k - \beta_k b_{k-1}$
  - Put $s := s + \xi_{k+1,n}\, b_{k+1}$
- Compute $u_n(x) = w_{v,\sigma^{-1}}(x)\, s$.

We note that in principle the special structure of the matrix $(a_{jk})_{j=1,k=0}^{n,\;\;n-1}$ would also allow us to construct a similar fast algorithm as described in the previous subsection. In contrast to that case the numerical tests showed, however, that this fast algorithm becomes extremely unstable for about $n \geq 50$. One possible explanation for this might be the fact that some of the sections $\mathbf{A}_m$ become nearly singular. Indeed it turns out that the corresponding numbers η_m defined in Lemma 8.2 get very small.

Hence, in the general case we made use of the recurrence relation for the columns only to compute the matrix entries efficiently and relied on Gauss elimination (with column pivoting) to solve the system.

Remark 8.8 *Finally, we point out that a possibility to maintain an $O(n^2)$-algorithm also in this general case is to proceed in a similar way as in the first case and write*

$$u_n(x) = w_{v,\sigma^{-1}}(x)\sum_{k=1}^{n} \xi_{kn}\frac{\widetilde{p}^v_n(x)}{(x - x^v_{kn})(\widetilde{p}^v_n)'(x^v_{kn})}.$$

Then one could express the $\widetilde{p}^v_n$ as a linear combination of the $\widetilde{p}^w_k$, $k = 0,\ldots,n$ ($w = w_{v,\sigma^{-1}}$) in order to be able to compute ϱ^w_n. We have not implemented this procedure, however.

9 A quadrature method for Cauchy singular integral equations with regular perturbations

We now consider perturbations of the Cauchy singular integral operator $aI + bS$ by an integral operator with a continuous (or weighted continuous) kernel, that is, we have equations of the form

$$(A+K)u = f, \tag{9.1}$$

where $A = aI + bS$ and

$$(Ku)(x) = \int_{-1}^{1} k(t,x)u(t)\,\mathrm{d}t$$

with a function $k : (-1,1)\times(-1,1) \to \mathbb{C}$. In what follows we will use a quadrature method to solve (9.1) approximately.

First we assume that the system of linear equations arising from the discretization of the unperturbed equation is such that the values $\xi_{kn} = w_n(x_{kn}^v)$ are sought, where $u_n = w_{v,\sigma^{-1}} w_n$ with some polynomial $w_n \in \Pi_n$. That means, we either have the special case $\sigma = v^{-1}$ and use the implementation described in Subsection 8.1, or we have the general case and proceed as it was proposed in Remark 8.8. In all what follows we suppose that

$$w_{v,\sigma}^{-1}(t)\,k(t,x) \in \mathbf{C}([-1,1]^2). \tag{9.2}$$

For $u \in \mathbf{L}^2_\sigma$ we define the operators

$$(K_n u)(x) = \int_{-1}^{1} [w_{v,\sigma} L_n^v w_{v,\sigma}^{-1} k(\cdot,x)](t)u(t)\,\mathrm{d}t,$$

where the Lagrange interpolation operator L_n^v is applied with respect to the first argument of k. Our quadrature method consists in finding $u_n \in X_n$ such that

$$A_{n,K}u_n := \widetilde{L_n^\sigma}(A+K_n)P_n^\sigma u_n = \widetilde{L_n^\sigma} f. \tag{9.3}$$

Let us consider the system of linear equations that arises from (9.3). For $u_n = w_{v,\sigma^{-1}} w_n \in X_n$ we have

$$\begin{aligned}(K_n u_n)(x) &= \int_{-1}^{1} \left[w_{v,\sigma} L_n^v w_{v,\sigma}^{-1} k(\cdot,x)\right](t)u_n(t)\,\mathrm{d}t \\ &= \int_{-1}^{1} v(t)\left[L_n^v w_{v,\sigma}^{-1} k(\cdot,x)\right](t)w_n(t)\,\mathrm{d}t \\ &= \sum_{k=1}^{n} A_{kn}^v w_{v,\sigma}^{-1}(x_{kn}^v)k(x_{kn}^v,x)\xi_{kn}.\end{aligned}$$

Hence, the unknown ξ_{kn} are determined by the equations

$$\sum_{k=1}^{n}\left(\frac{a_{jk}}{(\widetilde{p}_n^v)'(x_{kn}^v)} + A_{kn}^v w_{v,\sigma}^{-1}(x_{kn}^v)k(x_{kn}^v,x_{jn}^v)\right)\xi_{kn} = f(x_{jn}^v), \quad j=1,\dots,n, \tag{9.4}$$

where the coefficients a_{jk} are defined by (8.2).

If we do not have the situation $\sigma = v^{-1}$ and establish the system of discrete equations for $u_n = w_{v,\sigma^{-1}} \sum_{k=0}^{n-1} \xi_{kn} \widetilde{p}_k^w \in X_n$ $(w = w_{v,\sigma^{-1}})$ in the form (8.5), we obtain the system of linear equations

$$\sum_{k=0}^{n-1} \left(a_{jk} + \sum_{s=1}^{n} A_{sn}^v w_{v,\sigma}^{-1}(x_{sn}^v) k(x_{sn}^v, x_{jn}^v) \widetilde{p}_k^w (x_{sn}^v) \right) \xi_{kn} = f(x_{jn}^v), \quad j = 1, \ldots, n$$

to determine the ξ_{kn}, where the a_{jk} are the same as in (8.5).

In the following we are going to investigate the stability of the approximate operators $\widetilde{L_n^\sigma}(A + K_n)P_n^\sigma$. We will do this again by means of Banach algebra arguments. Another possible approach could be the theory of compactly convergent operator sequences (cf. [52], Satz 55 and 60).

Lemma 9.1 *Under the assumption (9.2), the sequence* $\{\widetilde{L_n^\sigma} K_n P_n^\sigma\}$ *belongs to* $\mathcal{I}$.

Proof. First, it follows immediately from Arzelà-Ascoli's theorem that we have $K \in \mathcal{K}(\mathbf{L}_\sigma^2, \mathbf{C}[-1,1])$, since

$$\left| \int_{-1}^{1} k(t,x) u(t) \, dt \right| \le \left\| w_{v,\sigma}^{-1}(\cdot) k(\cdot,\cdot) \right\|_\infty \int_{-1}^{1} v(t) \, dt \, \|u\|_\sigma$$

and

$$\left| \int_{-1}^{1} (k(t,x) - k(t,y)) u(t) \, dt \right|$$

$$\le \sup_{-1 \le t \le 1} \left| w_{v,\sigma}^{-1}(t) k(t,x) - w_{v,\sigma}^{-1}(t) k(t,y) \right| \int_{-1}^{1} v(t) \, dt \, \|u\|_\sigma$$

for all $x, y \in [-1,1]$.

Hence, we have $\left\| \widetilde{L_n^\sigma} K P_n^\sigma - K \right\|_{\mathcal{L}(\mathbf{L}_\sigma^2)} \to 0$, $(n \to \infty)$. Furthermore, we show that $\|K_n P_n^\sigma - K P_n^\sigma\|_{\mathcal{L}(\mathbf{L}_\sigma^2, \mathbf{C})} \to 0$. To this end, let $u \in \mathbf{L}_\sigma^2$ and $P_n^\sigma u = w_{v,\sigma^{-1}} w_n$, where w_n is some polynomial of degree less than n. We recall that the set of all polynomials in two variables is dense in $\mathbf{C}([-1,1]^2)$. Denote by $p_n(t,x)$ the polynomial of degree less than n (in both variables) of best uniform approximation to $w_{v,\sigma}^{-1}(t) k(t,x)$. Due to the accuracy of the Gaussian quadrature rule we have

$$|(K_n P_n^\sigma u - K P_n^\sigma u)(x)|$$

$$= \left| \int_{-1}^{1} w_{v,\sigma}(t) \left([L_n^v w_{v,\sigma}^{-1} k(\cdot, x)](t) (P_n^\sigma u)(t) - w_{v,\sigma}^{-1}(t) k(t,x) (P_n^\sigma u)(t) \right) dt \right|$$

$$\le \left| \sum_{j=1}^{n} A_{jn}^v \left(w_{v,\sigma}^{-1}(x_{jn}^v) k(x_{jn}^v, x) w_n(x_{jn}^v) - p_n(x_{jn}^v, x) w_n(x_{jn}^v) \right) \right|$$

$$+ \left| \int_{-1}^{1} v(t) \left(p_n(t,x) w_n(t) - w_{v,\sigma}^{-1}(t) k(t,x) w_n(t) \right) dt \right|$$

$$\leq \text{const}\ \left\|w_{v,\sigma}^{-1}k(\cdot,\cdot)-p_n\right\|_\infty \left(\sqrt{\sum_{j=1}^{n} A_{jn}^v |w_n(x_{jn}^v)|^2} + \sqrt{\int_{-1}^{1} v(t)|w_n(t)|^2\,dt}\right)$$

$$\leq c_n \left\|P_n^\sigma u\right\|_\sigma \qquad \text{with } c_n \to 0 \quad (n \to \infty)$$

uniformly with respect to $x \in [-1,1]$. If we make use of the fact that $\widetilde{L_n^\sigma} \to I$ (strongly) in $\mathcal{L}(\mathbf{C}, \mathbf{L}_\sigma^2)$, we arrive at $\left\|\widetilde{L_n^\sigma}(K_n - K)P_n^\sigma\right\|_{\mathcal{L}(\mathbf{L}_\sigma^2)} \to 0$. Thus, we have

$$\left\|\widetilde{L_n^\sigma}K_nP_n^\sigma - K\right\|_{\mathcal{L}(\mathbf{L}_\sigma^2)} \leq \left\|\widetilde{L_n^\sigma}KP_n^\sigma - K\right\|_{\mathcal{L}(\mathbf{L}_\sigma^2)} + \left\|\widetilde{L_n^\sigma}(K_n - K)P_n^\sigma\right\|_{\mathcal{L}(\mathbf{L}_\sigma^2)} \to 0$$

as $n \to \infty$. This shows that the sequence $\{\widetilde{L_n^\sigma}K_nP_n^\sigma\}$ is of the form $\{P_n^\sigma KP_n^\sigma + C_n\}$, where $\|C_n\| \to 0$, and is therefore contained in $\mathcal{I}$. ∎

Theorem 9.2 *The sequence* $\{\widetilde{L_n^\sigma}(A + K_n)P_n^\sigma\}$ *is stable if and only if the sequence* $\{\widetilde{L_n^\sigma}AP_n^\sigma\}$ *is stable and the operator* $A + K$ *is invertible in* $\mathbf{L}_\sigma^2$.

Proof. Lemma 9.1 shows that $\{\widetilde{L_n^\sigma}(A + K)P_n^\sigma\} \in \mathcal{A}$. Hence, we can apply Theorem 3.12. First note that the invertibility of $A + K$ implies that A is Fredholm with index 0 and hence also invertible. Further, the proof of Lemma 9.1 shows that the relation $\mathcal{W}_1(\{\widetilde{L_n^\sigma}(A + K_n)P_n^\sigma\} + \mathcal{N}) = \mathcal{W}_1(\{\widetilde{L_n^\sigma}AP_n^\sigma\} + \mathcal{N}) = \widetilde{A}$ holds. Since, finally, $\{\widetilde{L_n^\sigma}(A + K_n)P_n^\sigma\} + \mathcal{I} = \{\widetilde{L_n^\sigma}AP_n^\sigma\} + \mathcal{I}$, the assertion follows. ∎

10 Two fast algorithms for the quadrature method

We assume that we have the situation $\sigma = v^{-1}$ considered in Subsection 8.1, which allows the application of a fast algorithm with $O(n^2)$ operations to the collocation method for the unperturbed Cauchy singular integral equation. The special structure of the system of linear equations arising in the collocation method is destroyed by the presence of a perturbation K of the above-defined type. Under suitable smoothness conditions on the kernel k, however, one can construct a fast algorithm for the quadrature method making use of certain mapping properties of the operators involved in the scale of weighted Sobolev-like spaces introduced in Section 7. This algorithm is still based on the fast solution of the pure collocation equations for the unperturbed problem (with the approximate operators $A_n := A_{n,L}$) and a correction step on a lower level of discretization. Thus, again a complexity of $O(n^2)$ is obtained.

The basic idea for both algorithms described in the following is the observation that under the assumptions $K \in \mathcal{L}(\mathbf{L}^2_\sigma, \widetilde{\mathbf{L}}^2_{\sigma,s})$ and $A^{-1} \in \mathcal{L}(\widetilde{\mathbf{L}}^2_{\sigma,s})$ we have for the solution u of (9.1)

$$u - A^{-1}f = -A^{-1}Ku \in \widetilde{\mathbf{L}}^2_{\sigma,s},$$

hence the Fourier coefficients of this difference are decreasing very fast. This shows that the Fourier coefficients with high indices of the solution u can be computed with sufficient accuracy by applying the pure collocation method to the unperturbed equation (the solution of which is just $A^{-1}f$). The Fourier coefficients with low indices are then obtained by a correction step on a low level of discretization.

10.1 The approach of Amosov

The idea of this approach is due to Amosov ([1], see also [4]). The two steps of this fast algorithm can be described as follows: We search an approximate solution $u_n \in \operatorname{im} P_n^\sigma$ of (9.1), where we split up its Fourier series in two parts:

$$u_n = P_m^\sigma u_n + Q_m^\sigma u_n,$$

where $0 < m < n$ and $Q_m^\sigma = I - P_m^\sigma$. The two components of this decomposition are determined in the following way:

Algorithm 10.1

1. **Collocation method for unperturbed equation:**
   Put $Q_m^\sigma u_n = Q_m^\sigma v_n$, where $A_n v_n = \widetilde{L_n^\sigma} f$.

2. **Correction step:**
   Put $P_m^\sigma u_n = w_m$, where $(A_m + \widetilde{L_m^\sigma} K_m) w_m = \widetilde{L_m^\sigma} f - \widetilde{L_m^\sigma} A Q_m^\sigma v_n$.

Under suitable assumptions we can show an optimal convergence order of the sequence u_n in the Sobolev scale such as in Theorem 7.19.

Remark 10.2 *Throughout the present and the following subsection we will assume that the right hand side f of equation (9.1) is an element of $\widetilde{\mathbf{L}}^2_{\sigma,s}$. In view of the relation $u = A^{-1}f - A^{-1}Ku$ together with the mapping properties of K and A^{-1} (cf. the assumptions of the following propositions and lemmas) this will ensure that the solution u belongs to the same space in all places where this is not explicitly part of the hypothesis.*

In the following lemma we consider the error made in the first step of the fast algorithm.

Proposition 10.3 (cf. [4], Lemma 6.2) *Let $\{A_{n,L}\}$ be stable. Furthermore, assume that $A \in \mathcal{L}(\widetilde{\mathbf{L}}^2_{\sigma,s})$, $A^{-1} \in \mathcal{L}(\widetilde{\mathbf{L}}^2_{\sigma,s+\delta})$, $N_{s+\delta} \in \mathbf{L}^1(-1,1)$ for some $\delta > 0$, where $N_{s+\delta}$ is defined as in Lemma 7.27, and $s > \frac{1}{2}$. Then for $t \leq s$ we have the estimation*

$$\|Q^\sigma_m u - Q^\sigma_m u_n\|_{t,\sim} \leq \text{const}\,(m^{t-s-\delta} + n^{t-s})\,\|u\|_{s,\sim}\,.$$

Proof. We decompose the expression under consideration as follows:

$$Q^\sigma_m u - Q^\sigma_m u_n = Q^\sigma_m(u - A^{-1}f) + Q^\sigma_m A^{-1}(f - \widetilde{L^\sigma_n} f) + Q^\sigma_m A^{-1}(\widetilde{L^\sigma_n} f - Av_n)$$

Now let us estimate the three terms separately.

We have $u - A^{-1}f = -A^{-1}Ku \in \widetilde{\mathbf{L}}^2_{\sigma,s+\delta}$ due to the assumptions, and

$$\|Q^\sigma_m(u - A^{-1}f)\|_{t,\sim} \leq m^{t-s-\delta}\,\|u - A^{-1}f\|_{s+\delta,\sim}$$

$$\leq \text{const}\, m^{t-s-\delta}\,\|A^{-1}K\|_{\mathcal{L}(\widetilde{\mathbf{L}}^2_{\sigma,s},\widetilde{\mathbf{L}}^2_{\sigma,s+\delta})}\,\|u\|_{s,\sim}$$

by virtue of Lemma 7.3. Further,

$$\left\|Q^\sigma_m A^{-1}(f - \widetilde{L^\sigma_n} f)\right\|_{t,\sim} \leq \text{const}\,\|A^{-1}\|_{\mathcal{L}(\widetilde{\mathbf{L}}^2_{\sigma,t})}\, n^{t-s}\,\|f\|_{s,\sim}$$

$$\leq \text{const}\,\|A^{-1}\|_{\mathcal{L}(\widetilde{\mathbf{L}}^2_{\sigma,t})}\, n^{t-s}\,\|A + K\|_{\mathcal{L}(\widetilde{\mathbf{L}}^2_{\sigma,s})}\,\|u\|_{s,\sim}\,.$$

(Note that $A^{-1} \in \mathcal{L}(\widetilde{\mathbf{L}}^2_{\sigma,r})$ for all $r \leq s + \delta$ due to the interpolation property of the Sobolev spaces.) Finally, we have

$$\left\|Q^\sigma_m A^{-1}(\widetilde{L^\sigma_n} f - Av_n)\right\|_{t,\sim} = \left\|Q^\sigma_m A^{-1}(\widetilde{L^\sigma_n} - I)Av_n\right\|_{t,\sim}$$

$$\leq \text{const}\,\|A^{-1}\|_{\mathcal{L}(\widetilde{\mathbf{L}}^2_{\sigma,t})}\, n^{t-s}\,\|A\|_{\mathcal{L}(\widetilde{\mathbf{L}}^2_{\sigma,s})}\,\|v_n\|_{s,\sim}$$

and remember that (see Theorem 7.19)

$$\|v_n\|_{s,\sim} \leq \|v_n - P^\sigma_n v\|_{s,\sim} + \|P^\sigma_n v\|_{s,\sim} \leq \text{const}\,\|v\|_{s,\sim}$$

$$\leq \text{const}\,\|A^{-1}\|_{\mathcal{L}(\widetilde{\mathbf{L}}^2_{\sigma,s})}\,\|f\|_{s,\sim}$$

$$\leq \text{const}\,\|A^{-1}\|_{\mathcal{L}(\widetilde{\mathbf{L}}^2_{\sigma,s})}\,\|A + K\|_{\mathcal{L}(\widetilde{\mathbf{L}}^2_{\sigma,s})}\,\|u\|_{s,\sim}\,,$$

where v is the solution of the unperturbed equation $Av = f$. ■

Lemma 10.4 (cf. [4], Proof of Lemma 4.4) *Assume that* $w_{v,\sigma}^{-1}k(\cdot,x) \in \mathbf{L}^2_{v,s}$ *uniformly with respect to* x*, that is,* $\left\|w_{v,\sigma}^{-1}k(\cdot,x)\right\|_{v,s} \leq \text{const}$ *for all* $x \in [-1,1]$*, where* $s > \frac{1}{2}$*. Then the estimation*

$$\left\|\widetilde{L_m^\sigma}(K-K_n)u\right\|_{t,\sim} \leq \text{const}\, m^t\, n^{-s}\, \|u\|_\sigma$$

is true for all $u \in \mathbf{L}^2_\sigma$ *and* $t \leq s$.

Proof. We have

$$\left\|\widetilde{L_m^\sigma}(K-K_n)u\right\|^2_{t,\sim} \leq m^{2t}\left\|\widetilde{L_m^\sigma}(K-K_n)u\right\|^2_\sigma = m^{2t}\left\|L_m^v w_{v,\sigma^{-1}}^{-1}(K-K_n)u\right\|^2_v$$

$$= m^{2t}\sum_{j=1}^m A_{jm}^v w_{v,\sigma^{-1}}^{-2}(x_{jm}^v)$$

$$\times\left|\int_{-1}^1 w_{v,\sigma}(t)\left(w_{v,\sigma}^{-1}(t)k(t,x_{jm}^v) - \left[L_n^v w_{v,\sigma}^{-1}k(\cdot,x_{jm}^v)\right](t)\right)u(t)\,\mathrm{d}t\right|^2.$$

If we apply Schwarz' inequality in $\mathbf{L}^2(-1,1)$, we can estimate the latter expression by

$$m^{2t}\sum_{j=1}^m A_{jm}^v w_{v,\sigma^{-1}}^{-2}(x_{jm}^v)\left\|w_{v,\sigma}^{-1}k(\cdot,x_{jm}^v) - L_n^v w_{v,\sigma}^{-1}k(\cdot,x_{jm}^v)\right\|_v^2 \|u\|_\sigma^2$$

$$\leq \text{const}\, m^{2t}\, n^{-2s}\, \|u\|_\sigma^2 \left\|w_{v,\sigma}^{-1}k(\cdot,x_{jm}^v)\right\|^2_{v,s}$$

according to Theorem 7.16. (Note that $\sum_{j=1}^m A_{jm}^v w_{v,\sigma^{-1}}^{-2}(x_{jm}^v) \leq \text{const}\int_{-1}^1 \sigma(t)\,dt$ in view of Lemma 4.9.) ■

Proposition 10.5 *Assume that* $\{A_{n,K}\}$ *is stable,* $A \in \mathcal{L}(\widetilde{\mathbf{L}}^2_{\sigma,r})$ *for some* $r > \frac{1}{2}$*. Furthermore, let* $w_{v,\sigma}^{-1}k(\cdot,x) \in \mathbf{L}^2_{v,s}$ *uniformly with respect to* $x \in [-1,1]$*,* $N_s \in \mathbf{L}^1(-1,1)$*, and* $s > \frac{1}{2}$*. Then the following error estimate of the quadrature method (9.3) is valid if* $u \in \widetilde{\mathbf{L}}^2_{\sigma,s}$*:*

$$\|u_n - u\|_{t,\sim} \leq \text{const}\, n^{t-s}\, \|u\|_{s,\sim}$$

for $0 \leq t \leq s$.

Proof. We have

$$\|P_n^\sigma u - u_n\|_{t,\sim} \leq n^t\, \|P_n^\sigma u - u_n\|_\sigma$$

$$\leq n^t \text{const}\left\|\widetilde{L_n^\sigma}(A+K_n)P_n^\sigma u - \widetilde{L_n^\sigma}(A+K)u\right\|_\sigma.$$

The terms containing the operator A are treated as in the proof of Theorem 7.19 (note that due to Theorem 9.2 also $\{A_{n,L}\}$ is stable), the remaining ones can be estimated as follows:

$$\left\|\widetilde{L_n^\sigma} K_n P_n^\sigma u - \widetilde{L_n^\sigma} K u\right\|_\sigma \le \left\|\widetilde{L_n^\sigma}(K_n - K)P_n^\sigma u\right\|_\sigma + \left\|\widetilde{L_n^\sigma} K(P_n^\sigma u - u)\right\|_\sigma$$

$$\le \text{const}\, n^{-s} \left\|P_n^\sigma u\right\|_{s,\sim} + \left\|\widetilde{L_n^\sigma}\right\|_{\mathcal{L}(\widetilde{\mathbf{L}}^2_{\sigma,s},\mathbf{L}^2_\sigma)} \|K\|_{\mathcal{L}(\mathbf{L}^2_\sigma,\widetilde{\mathbf{L}}^2_{\sigma,s})} (1+n)^{-s} \|u\|_{s,\sim} .$$

(Compare Lemma 7.27 and the proof of Lemma 10.4.) ■

Corollary 10.6 *If the assumptions of Proposition 10.5 are fulfilled and* $(A+K)^{-1} \in \mathcal{L}(\widetilde{\mathbf{L}}^2_{\sigma,s})$, *then we have*

$$\left\|\left(\widetilde{L_n^\sigma}(A+K_n)\right)^{-1} P_n^\sigma f - (A+K)^{-1} f\right\|_{t,\sim} \le \text{const}\, n^{t-s} \|f\|_{s,\sim} .$$

for $0 \le t \le s$ *and all* $f \in \widetilde{\mathbf{L}}^2_{\sigma,s}$.

Proof. Let $u_n \in \operatorname{im} P_n^\sigma$ be the solution of

$$\widetilde{L_n^\sigma}(A+K_n)u_n = P_n^\sigma f.$$

Then, with $u = (A+K)^{-1}f$, we have

$$\left\|\left(\widetilde{L_n^\sigma}(A+K_n)\right)^{-1} P_n^\sigma f - (A+K)^{-1} f\right\|_{t,\sim} \le \|u_n - P_n^\sigma u\|_{t,\sim} + \|Q_n^\sigma u\|_{t,\sim} ,$$

and $\|Q_n^\sigma u\|_{t,\sim} \le$ const $n^{t-s} \|(A+K)^{-1}\|_{\mathcal{L}(\widetilde{\mathbf{L}}^2_{\sigma,s})} \|f\|_{s,\sim}$. Further, we can estimate as follows:

$$\|P_n^\sigma u - u_n\|_{t,\sim} \le n^t \|P_n^\sigma u - u_n\|_\sigma$$

$$\le \text{const}\, n^t \left\|\widetilde{L_n^\sigma}(A+K_n)P_n^\sigma u - P_n^\sigma(A+K)u\right\|_\sigma$$

$$\le \text{const}\, n^t \left(\left\|\widetilde{L_n^\sigma}(A+K_n)P_n^\sigma u - \widetilde{L_n^\sigma}(A+K)u\right\|_\sigma + \left\|(\widetilde{L_n^\sigma} - P_n^\sigma)f\right\|_\sigma\right).$$

The first expression can be estimated according to the proof of Proposition 10.5 by const $n^{t-s} \|(A+K)^{-1}\|_{\mathcal{L}(\widetilde{\mathbf{L}}^2_{\sigma,s})} \|f\|_{s,\sim}$, the second one by

$$\text{const}\, n^t \left(\left\|(\widetilde{L_n^\sigma} - I)f\right\|_\sigma + \|Q_n^\sigma f\|_\sigma\right) \le \text{const}\, n^{t-s} \|f\|_{s,\sim} ,$$

and we have proved the Corollary. ■

Corollary 10.7 *Under the assumptions of Corollary 10.6 the sequence* $\{\widetilde{L_n^\sigma}(A+K_n)P_n^\sigma\}$ *is stable in* $\widetilde{\mathbf{L}}^2_{\sigma,t}$ *for all* $t \le s$.

Proof. Corollary 10.6 immediately gives the stability in $\widetilde{\mathbf{L}}^2_{\sigma,s}$. For $t < s$, the assertion follows from the interpolation property. ■

Now we are going to give an error estimate for the second step of our algorithm.

Proposition 10.8 (cf. [4], Lemma 6.4) *Assume that the sequence $\{A_{n,K}\}$ is stable in $\mathbf{L}^2_\sigma$, $A \in \mathcal{L}(\widetilde{\mathbf{L}}^2_{\sigma,s})$, $A^{-1} \in \mathcal{L}(\widetilde{\mathbf{L}}^2_{\sigma,s+\delta})$, $N_{s+\delta} \in \mathbf{L}^1$, where $s > \frac{1}{2}$, $\delta > 0$. Furthermore, let $(A+K)^{-1} \in \mathcal{L}(\widetilde{\mathbf{L}}^2_{\sigma,s})$, and let, finally, $w_{v,\sigma}^{-1}k(\cdot,x) \in \mathbf{L}^2_{v,s+\delta}$ uniformly with respect to x. Then the estimation*

$$\|P^\sigma_m u - P^\sigma_m u_n\|_{t,\sim} \le \text{const}\,(m^{t-s-\delta} + n^{t-s})\,\|u\|_{s,\sim}$$

holds for $\frac{1}{2} < t \le s$.

Proof. By definition, we have $P^\sigma_m u_n = w_m$, where $\widetilde{L^\sigma_m}(A+K_m)w_m = \widetilde{L^\sigma_m} f - \widetilde{L^\sigma_m} A Q^\sigma_m v_n$. Since, by virtue of Corollary 10.7, the sequence $\{\widetilde{L^\sigma_m}(A+K_m)P^\sigma_m\}$ is stable in $\widetilde{\mathbf{L}}^2_{\sigma,t}$, it is sufficient to estimate $\left\|\widetilde{L^\sigma_m}(A+K_m)(w_m - P^\sigma_m u)\right\|_{t,\sim}$. Thus, we have to consider

$$\begin{aligned}
&\widetilde{L^\sigma_m}(A+K_m)(w_m - P^\sigma_m u)\\
&= \widetilde{L^\sigma_m} f - \widetilde{L^\sigma_m} A Q^\sigma_m u_n - \widetilde{L^\sigma_m}(A+K_m)P^\sigma_m u\\
&= \widetilde{L^\sigma_m}(A+K)(P^\sigma_m u + Q^\sigma_m u) - \widetilde{L^\sigma_m} A Q^\sigma_m u_n - \widetilde{L^\sigma_m}(A P^\sigma_m + K_m P^\sigma_m)u\\
&= \widetilde{L^\sigma_m}(K-K_m)u + \widetilde{L^\sigma_m} A(Q^\sigma_m u - Q^\sigma_m u_n) + \widetilde{L^\sigma_m} K_m Q^\sigma_m u.
\end{aligned}$$

The first term can be estimated by

$$\left\|\widetilde{L^\sigma_m}(K-K_m)u\right\|_{t,\sim} \le \text{const}\, m^{t-s-\delta}\,\|u\|_\sigma$$

in view of Lemma 10.4. For the second one, we have

$$\left\|\widetilde{L^\sigma_m} A(Q^\sigma_m u - Q^\sigma_m u_n)\right\|_{t,\sim} \le \text{const}\,\left\|\widetilde{L^\sigma_m} A\right\|_{\mathcal{L}(\widetilde{\mathbf{L}}^2_{\sigma,t})} (m^{t-s-\delta} + n^{t-s})\,\|u\|_{s,\sim}$$

according to Proposition 10.3. Finally, we can show that the third term vanishes, since we have

$$\begin{aligned}
(K_m Q^\sigma_m u)(x) &= \int_{-1}^{1} \left[w_{v,\sigma} L^v_m w_{v,\sigma}^{-1} k(\cdot,x)\right](t)(Q^\sigma_m u)(t)\,\mathrm{d}t\\
&= \sum_{j=1}^{m} w_{v,\sigma}^{-1}(x^v_{jm})k(x^v_{jm},x)\int_{-1}^{1} w_{v,\sigma}(t) l^v_{jm}(t)(Q^\sigma_m u)(t)\,\mathrm{d}t\\
&= \sum_{j=1}^{m} w_{v,\sigma}^{-1}(x^v_{jm})k(x^v_{jm},x)\langle w_{v,\sigma^{-1}} l^v_{jm}, Q^\sigma_m u\rangle_\sigma = 0
\end{aligned}$$

because $w_{v,\sigma^{-1}} l^v_{jm} \in \operatorname{im} P^\sigma_m$, where l^v_{jm} denote the fundamental polynomials of Lagrange interpolation. ■

Remark 10.9 *The assumption on the boundedness of* $(A+K)^{-1}$ *is only needed to guarantee the stability of* $\{\widetilde{L_n^\sigma}(A+K_n)P_n^\sigma\}$ *in* $\widetilde{\mathbf{L}}^2_{\sigma,t}$. *Thus, in order to get the estimation in Proposition 10.8 for some fixed* $t<s$, *it is actually sufficient to require* $(A+K)^{-1}\in\mathcal{L}(\widetilde{\mathbf{L}}^2_{\sigma,r})$ *for some* $r>\frac{1}{2}$, $r\geq t$.

Remark 10.10 *If we dispense with the assumption* $t>\frac{1}{2}$ *in Proposition 10.8, we can get the estimation*

$$\|P_m^\sigma u - P_m^\sigma u_n\|_{t,\sim} \leq \text{const}\,(m^{t-s-\delta}+n^{t-s}+m^{t-t'}n^{t'-s})\,\|u\|_{s,\sim}\,,$$

where t' *is an arbitrary number satisfying* $t'>\frac{1}{2}$ *and* $t\leq t'\leq s$.

Proof. We decompose the expression to be estimated as in the proof of Proposition 10.8 and estimate the second term occurring there as follows:

$$\begin{aligned}
&\left\|\widetilde{L_m^\sigma}A(Q_m^\sigma u - Q_m^\sigma u_n)\right\|_{t,\sim}\\
&\quad\leq \left\|(\widetilde{L_m^\sigma}-I)A(Q_m^\sigma u - Q_m^\sigma u_n)\right\|_{t,\sim} + \|A(Q_m^\sigma u - Q_m^\sigma u_n)\|_{t,\sim}\\
&\quad\leq \text{const}\, m^{t-t'}\,\|A\|_{\mathcal{L}(\widetilde{\mathbf{L}}^2_{\sigma,t'})}\,\|Q_m^\sigma u - Q_m^\sigma u_n\|_{t',\sim} + \|A\|_{\mathcal{L}(\widetilde{\mathbf{L}}^2_{\sigma,t})}\,\|Q_m^\sigma u - Q_m^\sigma u_n\|_{t,\sim}\\
&\quad\leq \text{const}\, m^{t-t'}(m^{t'-s-\delta}+n^{t'-s})\,\|u\|_{s,\sim} + \text{const}\,(m^{t-s-\delta}+n^{t-s})\,\|u\|_{s,\sim}\,.
\end{aligned}$$

The remaining terms can be treated in the same way as before. ■

We are now in a position to summarize our results on the fast algorithm in the following theorem.

Theorem 10.11 *Let the assumptions of Proposition 10.8 be satisfied with* $\delta\geq\frac{s-t}{2}$, *where* $\frac{1}{2}<t\leq s$. *If we choose the order* m *of the system in the correction step such that* $m\sim n^{2/3}$, *we can obtain an approximate solution* $u_n\in\operatorname{im}P_n^\sigma$ *of equation (9.1) that satisfies the error estimate*

$$\|u_n-u\|_{t,\sim}\leq \text{const}\, n^{t-s}\,\|u\|_{s,\sim}$$

with a computational complexity of $O(n^2)$ *operations.*

Proof. Since we can solve the system of the collocation equations for the unperturbed equation with $O(n^2)$ operations, the total computational complexity remains of order $O(n^2)$ if we choose m such that $m\sim n^{2/3}$, that is,

$$c_1 n^2 \leq m^3 \leq c_2 n^2$$

with some constants $c_1, c_2>0$.

As for the error estimate, we have

$$\begin{aligned}
\|u_n-u\|^2_{t,\sim} &= \|P_m^\sigma u_n - P_m^\sigma u\|^2_{t,\sim} + \|Q_m^\sigma u_n - Q_m^\sigma u\|^2_{t,\sim}\\
&\leq \text{const}\,\left(m^{t-s-\delta}+n^{t-s}\right)^2\|u\|^2_{s,\sim} \leq \text{const}\,\left(n^{\frac{2}{3}(t-s-\delta)}+n^{t-s}\right)^2\|u\|^2_{s,\sim}\\
&\leq \text{const}\,n^{2(t-s)}\,\|u\|^2_{s,\sim}\,. \quad ■
\end{aligned}$$

Remark 10.12 *If $t \le \frac{1}{2}$ and the remaining assumptions of Theorem 10.11 are satisfied, the error estimate is weakened to*

$$\|u_n - u\|_{t,\sim} \le \text{const}\, n^{t-s+(t'-t)/3}\, \|u\|_{s,\sim},$$

where $t' > \frac{1}{2}$ is arbitrary.

Proof. Use Remark 10.10, where the additional term occurring there is estimated by

$$m^{t-t'} n^{t'-s} \le \text{const}\, n^{\frac{2}{3}(t-t')} n^{t'-s} = \text{const}\, n^{t-s+(t'-t)/3}. \quad \blacksquare$$

Remark 10.13 *The Amosov-type algorithm requires several transformations between function values and Fourier coefficients of weighted polynomials, but the complexity of these procedures does not exceed $O(n^2)$ (see Algorithms 8.5 and 8.7).*

Let us have a look at two examples. Both are considered for the case $v = \varphi$, $\sigma = \varphi^{-1}$.

(K1) $a(x) = 2, \qquad b(x) = i(1-x^2)^{3.1}, \qquad k(t,x) = (1 + x\sin t)(1-x^2)^{3/2},$

$$u(x) = |x|\sqrt{1-x^2}$$

(K2) a, b, k as in (K1), $\qquad u(x) = 1 - x^2$

Example (K1)				
n	m	$\varepsilon_{n,0}$	$n\,\varepsilon_{n,0}$	$n^{1/2}\varepsilon_{n,1}$
100	13	0.004246790769800	0.424679	1.266984
200	21	0.001942214086338	0.388443	1.271408
400	33	0.000914213702974	0.365685	1.278040
800	52	0.000385658023868	0.308526	1.292847
1000	60	0.000309956369540	0.309956	1.293685
2000	96	0.000148183772967	0.296368	1.283635
4000	152	0.000072018155384	0.288073	1.252454
6000	199	0.000058321528277	0.349929	1.267474
8000	240	0.000035991256976	0.287930	1.179145
10000	279	0.000035052300938	0.350523	1.206333
12000	315	0.000029195287536	0.350343	1.169947
14000	349	0.000025022917736	0.350321	1.130465
16000	381	0.000021928506543	0.350856	1.088310

In both cases we have $A, A^{-1} \in \mathcal{L}(\widetilde{\mathbf{L}}^2_{\sigma,s})$ for $s = 3$. Further, we have $k(\cdot, x) \in \mathbf{L}^2_{\varphi,s}$ for arbitrary s uniformly with respect to x and $N_s \in \mathbf{L}^1$ at least for $s = 3$.

In Example (K1) we have $u \in \widetilde{\mathbf{L}}^2_{\sigma,s}$ for $s < \frac{3}{2}$. Thus, if we assume that the conditions on the mapping properties of $(A+K)^{-1}$ in Remark 10.9 are satisfied (which seems difficult to verify), Theorem 10.11 would yield convergence rates of

$$\varepsilon_{n,0} \le \varepsilon_{n,1/2+\eta} \le \text{const}\, n^{-1+\eta}, \qquad \varepsilon_{n,1} \le \text{const}\, n^{-1/2+\eta},$$

where $\eta > 0$ is arbitrarily small. This is confirmed by the numerical results. On the other hand, the results in this case seem to suggest that the estimation for $\varepsilon_{n,0}$ can not be improved, that is, we could not observe the somewhat better convergence rate $\varepsilon_{n,0} \leq \text{const}\, n^{-\frac{4}{3}+\eta}$, $\eta > 0$, which would follow from Remark 10.12. A possible explanation for this observation might be the fact that the constant in this estimation can be extremely large if t' in Remark 10.12 is chosen near $\frac{1}{2}$, since it contains $\left\|\widetilde{L_n^\sigma} - I\right\|_{\mathcal{L}(\widetilde{\mathbf{L}}^2_{\sigma,t'},\widetilde{\mathbf{L}}^2_{\sigma,t})}$.

In Example (K2), we have $u \in \widetilde{\mathbf{L}}^2_{\sigma,s}$, $s < \frac{5}{2}$. This would suggest the estimation

$$\varepsilon_{n,0} \leq \varepsilon_{n,1/2} \leq \text{const}\, n^{-2+\eta}, \qquad \varepsilon_{n,3/2} \leq \text{const}\, n^{-1+\eta}$$

for arbitrarily small $\eta < 0$ (where, however, for the first relation not all conditions of Theorem 10.11 can be verified). We see that the convergence is not monotonic, one can observe a different behaviour for the subsequences where m is even or odd, respectively. To illustrate that the predicted convergence rates are obtained nevertheless, we present the results for more n here than in the previous example.

Example (K2)				
n	m	$\varepsilon_{n,0}$	$n^2\,\varepsilon_{n,0}$	$n\,\varepsilon_{n,3/2}$
100	13	0.000174661430412	1.746614	1.974791
200	21	0.000002899204247	0.115968	1.751942
400	33	0.000000504646877	0.080744	1.755752
800	52	0.000002480272778	1.587375	1.903416
1000	60	0.000001613666765	1.613667	1.906060
2000	96	0.000000393391951	1.573568	1.899504
4000	152	0.000000099008872	1.584142	1.881441
5000	176	0.000000063761805	1.594045	1.867334
6000	199	0.000000000542838	0.019542	1.693618
7000	220	0.000000032635824	1.599155	1.826785
8000	240	0.000000025135204	1.608653	1.801483
9000	260	0.000000019767769	1.601189	1.770559
10000	279	0.000000000148731	0.014873	1.568862
11000	297	0.000000000116465	0.014092	1.525686
12000	315	0.000000000092861	0.013372	1.476933
13000	332	0.000000009492065	1.604159	1.605196
14000	349	0.000000000061585	0.012071	1.360327
15000	365	0.000000000050862	0.011444	1.290692
16000	381	0.000000000042207	0.010805	1.211924
17000	397	0.000000000035063	0.010133	1.122822
18000	412	0.000000004966040	1.608997	1.262221
19000	427	0.000000000023832	0.008603	0.895882
20000	442	0.000000004021710	1.608684	1.055039

10.2 A fixed point iteration method

In this subsection, we investigate a second type of fast algorithms for the efficient solution of (9.1) that makes use of a fixed point iteration method. It is based on the ideas of [5].

For fixed positive integers $0 < m < n$ we choose $u_{n,0} \in \text{im } P_n^\sigma$ arbitrary and put

$$u_{n,j+1} := v_{n,j} + w_{m,n,j}, \qquad j = 0, 1, 2, \ldots,$$

where $w_{m,n,j} \in \text{im } P_m^\sigma$ is determined by

$$\widetilde{L_m^\sigma}(A + K_m) w_{m,n,j} = \widetilde{L_m^\sigma} K_n (u_{n,j} - v_{n,j})$$

and $v_{n,j} \in \text{im } P_n^\sigma$ is the solution of

$$A_n v_{n,j} = \widetilde{L_n^\sigma} f - \widetilde{L_n^\sigma} K_n u_{n,j}.$$

Lemma 10.14 ([5]) *The algorithm formulated above can be described by the iteration process*

$$u_{n,j+1} = T_{mn} \widetilde{L_n^\sigma} f + (P_n^\sigma - T_{mn} \widetilde{L_n^\sigma}(A + K_n)) u_{n,j}, \tag{10.1}$$

where

$$T_{mn} = (P_n^\sigma - [\widetilde{L_m^\sigma}(A + K_m) P_m^\sigma]^{-1} \widetilde{L_m^\sigma} K_n) A_n^{-1} P_n^\sigma.$$

The solution u_n of (9.3) is a fixed point of this iteration process.

Proof. We have

$$\begin{aligned}
u_{n,j+1} &= A_n^{-1} \widetilde{L_n^\sigma} f - A_n^{-1} \widetilde{L_n^\sigma} K_n u_{n,j} + [\widetilde{L_m^\sigma}(A + K_m) P_m^\sigma]^{-1} \widetilde{L_m^\sigma} K_n u_{n,j} \\
&\quad - [\widetilde{L_m^\sigma}(A + K_m) P_m^\sigma]^{-1} \widetilde{L_m^\sigma} K_n A_n^{-1} \widetilde{L_n^\sigma} f \\
&\quad + [\widetilde{L_m^\sigma}(A + K_m) P_m^\sigma]^{-1} \widetilde{L_m^\sigma} K_n A_n^{-1} \widetilde{L_n^\sigma} K_n u_{n,j} \\
&= A_n^{-1} \widetilde{L_n^\sigma} f - [\widetilde{L_m^\sigma}(A + K_m) P_m^\sigma]^{-1} \widetilde{L_m^\sigma} K_n A_n^{-1} \widetilde{L_n^\sigma} f + u_{n,j} \\
&\quad - A_n^{-1} \widetilde{L_n^\sigma}(A + K_n) u_{n,j} + [\widetilde{L_m^\sigma}(A + K_m) P_m^\sigma]^{-1} \widetilde{L_m^\sigma} K_n u_{n,j} \\
&\quad + [\widetilde{L_m^\sigma}(A + K_m) P_m^\sigma]^{-1} \widetilde{L_m^\sigma} K_n A_n^{-1} \widetilde{L_n^\sigma} K_n u_{n,j} \\
&= T_{mn} \widetilde{L_n^\sigma} f + (P_n^\sigma - T_{mn} \widetilde{L_n^\sigma}(A + K_n)) u_{n,j}.
\end{aligned}$$

If now $\widetilde{L_n^\sigma}(A + K_n) u_n = \widetilde{L_n^\sigma} f$, the relation

$$u_n = T_{mn} \widetilde{L_n^\sigma} f + (P_n^\sigma - T_{mn} \widetilde{L_n^\sigma}(A + K_n)) u_n \tag{10.2}$$

immediately follows. ∎

Thus, for proving the convergence of our algorithm, we have to show that

$$\|P_n^\sigma - T_{mn}\widetilde{L_n^\sigma}(A+K_n)P_n^\sigma\|_{\mathcal{L}(\mathbf{L}^2_\sigma)} < 1$$

for sufficiently large n and m. Under this condition, there exists a unique fixed point, which shows that (10.2) and (9.3) are equivalent.

The iteration formula (10.1) can also be motivated by the following arguments: Provided that an approximate solution $u_{n,j}$ of (9.3) is given, we would determine the exact solution by $u_{n,j+1} := u_{n,j} + d_{n,j}$, where $d_{n,j}$ satisfies the defect equation

$$\widetilde{L_n^\sigma}(A+K_n)d_{n,j} = \widetilde{L_n^\sigma}f - \widetilde{L_n^\sigma}(A+K_n)u_{n,j},$$

or equivalently,

$$d_{n,j} = [\widetilde{L_n^\sigma}(A+K_n)P_n^\sigma]^{-1}(\widetilde{L_n^\sigma}f - \widetilde{L_n^\sigma}(A+K_n)u_{n,j}).$$

It is easy to see that

$$\begin{aligned}[\widetilde{L_n^\sigma}(A+K_n)P_n^\sigma]^{-1} &= (P_n^\sigma + A_n^{-1}\widetilde{L_n^\sigma}K_nP_n^\sigma)^{-1}A_n^{-1}P_n^\sigma \\ &= (P_n^\sigma - [\widetilde{L_n^\sigma}(A+K_n)P_n^\sigma]^{-1}\widetilde{L_n^\sigma}K_nP_n^\sigma)A_n^{-1}P_n^\sigma.\end{aligned}$$

We arrive at (10.1) if we replace $[\widetilde{L_n^\sigma}(A+K_n)P_n^\sigma]^{-1}\widetilde{L_n^\sigma}$ by $[\widetilde{L_m^\sigma}(A+K_m)P_m^\sigma]^{-1}\widetilde{L_m^\sigma}$, that is, we solve the perturbed equation occurring in the iteration process on a lower level of discretization.

The following lemma gives a decomposition of $P_n^\sigma - T_{mn}\widetilde{L_n^\sigma}(A+K_n)P_n^\sigma$ that will enable us to estimate the norm of this operator.

Lemma 10.15 ([5]) *The operator $P_n^\sigma - T_{mn}\widetilde{L_n^\sigma}(A+K_n)P_n^\sigma$ admits the representation*

$$\begin{aligned}&[\widetilde{L_m^\sigma}(A+K_m)P_m^\sigma]^{-1}\Big[(\widetilde{L_m^\sigma}K_n - \widetilde{L_m^\sigma}K_m)A_n^{-1}\widetilde{L_n^\sigma}K_nP_n^\sigma \\ &\quad +\widetilde{L_m^\sigma}K_m(A_n^{-1}\widetilde{L_n^\sigma}K_n - A_m^{-1}\widetilde{L_m^\sigma}K_n)\Big] + (A_m^{-1}\widetilde{L_m^\sigma}K_n - A_n^{-1}\widetilde{L_n^\sigma}K_n)P_n^\sigma.\end{aligned}$$

Proof. We have

$$\begin{aligned}&P_n^\sigma - T_{mn}\widetilde{L_n^\sigma}(A+K_n)P_n^\sigma \\ &\quad= -A_n^{-1}\widetilde{L_n^\sigma}K_nP_n^\sigma + [\widetilde{L_m^\sigma}(A+K_m)P_m^\sigma]^{-1}\widetilde{L_m^\sigma}K_nP_n^\sigma \\ &\qquad +[\widetilde{L_m^\sigma}(A+K_m)P_m^\sigma]^{-1}\widetilde{L_m^\sigma}K_nA_n^{-1}\widetilde{L_n^\sigma}K_nP_n^\sigma \\ &\quad= -A_n^{-1}\widetilde{L_n^\sigma}K_nP_n^\sigma + [\widetilde{L_m^\sigma}(A+K_m)P_m^\sigma]^{-1}\widetilde{L_m^\sigma}K_nA_n^{-1}\widetilde{L_n^\sigma}K_nP_n^\sigma \\ &\qquad -[\widetilde{L_m^\sigma}(A+K_m)P_m^\sigma]^{-1}\widetilde{L_m^\sigma}K_mA_m^{-1}\widetilde{L_m^\sigma}K_nP_n^\sigma + A_m^{-1}\widetilde{L_m^\sigma}K_nP_n^\sigma,\end{aligned}$$

and it is easy to see that this equals the expression given in the assertion. ■

Now we are in a position to show the convergence of (10.1).

Theorem 10.16 *Assume that* $\{A_{n,K}\}$ *is stable (in* $\mathbf{L}^2_\sigma$*),* $A \in \mathcal{L}(\widetilde{\mathbf{L}}^2_{\sigma,r})$ *for some* $r > \frac{1}{2}$*,* $A^{-1}, (A+K)^{-1} \in \mathcal{L}(\widetilde{\mathbf{L}}^2_{\sigma,s})$*,* $s > \frac{1}{2}$*,* $N_s \in \mathbf{L}^1$ *and* $\left\|w_{v,\sigma}^{-1} k(\cdot,x)\right\|_{v,s} \le \text{const}$ *for all* $x \in (-1,1)$*. Then the iteration process (10.1) converges to the solution* u_n *of (9.3), and we have the error estimate*

$$\|u_n - u_{n,j}\|_{t,\sim} \le \text{const}\, m^{j(t-s)}\, \|u_n - u_{n,0}\|_{t,\sim} \qquad \textit{for all} \quad 0 \le t < s.$$

Proof. We will show that $\left\|P_n^\sigma - T_{mn}\widetilde{L_n^\sigma}(A+K_n)P_n^\sigma\right\|_{\mathcal{L}(\widetilde{\mathbf{L}}^2_{\sigma,t})} \le \text{const}\, m^{t-s}$ for sufficiently large n and m, where we make use of the decomposition proved in Lemma 10.15.

First we note that $\|A_n^{-1}P_n^\sigma\|_{\mathcal{L}(\widetilde{\mathbf{L}}^2_{\sigma,t})}, \left\|(\widetilde{L_m^\sigma}(A+K_m)P_m^\sigma)^{-1}\right\|_{\mathcal{L}(\widetilde{\mathbf{L}}^2_{\sigma,t})} \le \text{const}$ due to Corollary 10.7 (for the first norm, consider the special case $k(t,x) \equiv 0$).

Moreover, we have

$$\left\|\widetilde{L_n^\sigma}K_n\right\|_{\mathcal{L}(\widetilde{\mathbf{L}}^2_{\sigma,t},\widetilde{\mathbf{L}}^2_{\sigma,s})} \le \left\|\widetilde{L_n^\sigma}(K_n-K)\right\|_{\mathcal{L}(\widetilde{\mathbf{L}}^2_{\sigma,t},\widetilde{\mathbf{L}}^2_{\sigma,s})} + \left\|\widetilde{L_n^\sigma}\right\|_{\mathcal{L}(\widetilde{\mathbf{L}}^2_{\sigma,s})} \|K\|_{\mathcal{L}(\widetilde{\mathbf{L}}^2_{\sigma,t},\widetilde{\mathbf{L}}^2_{\sigma,s})} \le \text{const}$$

in view of Lemmas 7.27 and 10.4. From Lemma 10.4 we can further conclude

$$\left\|\widetilde{L_m^\sigma}K_n - \widetilde{L_m^\sigma}K_m\right\|_{\mathcal{L}(\widetilde{\mathbf{L}}^2_{\sigma,t})} \le \left\|\widetilde{L_m^\sigma}(K_n-K)\right\|_{\mathcal{L}(\widetilde{\mathbf{L}}^2_{\sigma,t})}$$

$$+ \left\|\widetilde{L_m^\sigma}(K_m-K)\right\|_{\mathcal{L}(\widetilde{\mathbf{L}}^2_{\sigma,t})} \le \text{const}\,(m^t n^{-s} + m^{t-s}).$$

If we note that

$$\left\|\widetilde{L_n^\sigma}K_n - \widetilde{L_m^\sigma}K_n\right\|_{\mathcal{L}(\widetilde{\mathbf{L}}^2_{\sigma,t})} \le \left\|\widetilde{L_n^\sigma}(K_n-K)\right\|_{\mathcal{L}(\widetilde{\mathbf{L}}^2_{\sigma,t})} + \left\|\widetilde{L_m^\sigma}(K_n-K)\right\|_{\mathcal{L}(\widetilde{\mathbf{L}}^2_{\sigma,t})}$$

$$+ \left\|\widetilde{L_n^\sigma} - I\right\|_{\mathcal{L}(\widetilde{\mathbf{L}}^2_{\sigma,s},\widetilde{\mathbf{L}}^2_{\sigma,t})} \|K\|_{\mathcal{L}(\widetilde{\mathbf{L}}^2_{\sigma,t},\widetilde{\mathbf{L}}^2_{\sigma,s})} + \left\|\widetilde{L_m^\sigma} - I\right\|_{\mathcal{L}(\widetilde{\mathbf{L}}^2_{\sigma,s},\widetilde{\mathbf{L}}^2_{\sigma,t})} \|K\|_{\mathcal{L}(\widetilde{\mathbf{L}}^2_{\sigma,t},\widetilde{\mathbf{L}}^2_{\sigma,s})}$$

$$\le \text{const}\,(m^{t-s} + n^{t-s}),$$

we can finally estimate (cf. Corollary 10.6 with $k(t,x) \equiv 0$)

$$\left\|A_n^{-1}\widetilde{L_n^\sigma}K_n - A_m^{-1}\widetilde{L_m^\sigma}K_n\right\|_{\mathcal{L}(\widetilde{\mathbf{L}}^2_{\sigma,t})} \le \|A_n^{-1}P_n^\sigma\|_{\mathcal{L}(\widetilde{\mathbf{L}}^2_{\sigma,t})} \left\|\widetilde{L_n^\sigma}K_n - \widetilde{L_m^\sigma}K_n\right\|_{\mathcal{L}(\widetilde{\mathbf{L}}^2_{\sigma,t})}$$

$$+ \|A_n^{-1}P_n^\sigma - A^{-1}\|_{\mathcal{L}(\widetilde{\mathbf{L}}^2_{\sigma,s},\widetilde{\mathbf{L}}^2_{\sigma,t})} \left\|\widetilde{L_m^\sigma}K_n\right\|_{\mathcal{L}(\widetilde{\mathbf{L}}^2_{\sigma,t},\widetilde{\mathbf{L}}^2_{\sigma,s})}$$

$$+ \|A^{-1} - A_m^{-1}P_m^\sigma\|_{\mathcal{L}(\widetilde{\mathbf{L}}^2_{\sigma,s},\widetilde{\mathbf{L}}^2_{\sigma,t})} \left\|\widetilde{L_m^\sigma}K_n\right\|_{\mathcal{L}(\widetilde{\mathbf{L}}^2_{\sigma,t},\widetilde{\mathbf{L}}^2_{\sigma,s})} \le \text{const}\,(n^{t-s} + m^{t-s}).$$

Thus, keeping in mind Lemma 10.15, we have arrived at the norm estimation $\left\|P_n^\sigma - T_{mn}\widetilde{L_n^\sigma}(A+K_n)P_n^\sigma\right\|_{\mathcal{L}(\widetilde{\mathbf{L}}^2_{\sigma,t})} \le \text{const}\, m^{t-s}$, and the convergence of (10.1) as well as the error estimate for $\|u_n - u_{n,j}\|_{t,\sim}$ are now an immediate consequence of Banach's fixed point theorem. ■

Remark 10.17 *With respect to the boundedness of* $(A+K)^{-1}$ *and* A^{-1} *we can refer to Remark 10.9.*

Corollary 10.18 *Under the assumptions of the preceding theorem, one can apply the iteration process (10.1) to (9.3) obtaining an error that is of the same order as the discretization error* $\|u-u_n\|_{t,\sim} \leq \text{const } n^{t-s} \|u\|_{s,\sim}$ *(according to Proposition 10.5) with a computational complexity of* $O(n^2)$ *operations.*

Proof. We specify that we always choose $u_{n,0}=0$, whence

$$\|u_{n,j}-u_n\|_{t,\sim} \leq \text{const } m^{j(t-s)} \|u_n\|_{t,\sim} \leq \text{const } m^{j(t-s)} \|u\|_{s,\sim}$$

(cf. Proposition 10.5). Further, we choose the lower discretization level m such that

$$c_1\, n^{1/2} \leq m \leq c_2\, n^{2/3}$$

with some constants $c_1, c_2 > 0$. Then the iteration method requires $O(n^2+m^3)=O(n^2)$ operations per step. If we always perform 2 iteration steps, this results in an error of the order

$$\|u_{n,j}-u_n\|_{t,\sim} \leq \text{const } n^{t-s} \|u\|_{s,\sim}. \quad ■$$

One advantage of this iteration method in comparison with that one described in the preceding subsection is the fact that the assumptions on the smoothness of the kernel k and on the boundedness of the operator A^{-1} in Sobolev spaces (cf. the propositions 10.3 and 10.8) are slightly weaker. A second advantage is that the convergence rate of n^{t-s} remains valid for all $0 \leq t < s$, whereas in the Amosov-type algorithm there is a loss if $t \leq \frac{1}{2}$.

It could be considered as a disadvantage that one has to carry out two iteration steps, but the asymptotic order of the computational complexity remains the same. Besides, the order of the small system can be chosen lower.

We conclude our considerations by providing numerical results for the examples (K1) and (K2) from above. They show that the loss in the case $t < \frac{1}{2}$ does not appear.

Example (K1)				
n	m	$\varepsilon_{n,0}$	$n^{3/2}\,\varepsilon_{n,0}$	$n^{1/2}\varepsilon_{n,1}$
100	10	0.0009010	0.901	1.216
200	15	0.0003193	0.903	1.212
400	20	0.0001130	0.904	1.207
800	29	0.0000400	0.905	1.196
1000	32	0.0000286	0.905	1.191
2000	45	0.0000101	0.905	1.163
4000	64	0.0000035	0.904	1.107
6000	78	0.0000019	0.899	1.048
8000	90	0.0000012	0.890	0.986
10000	100	0.0000008	0.876	0.919

Example (K2)				
n	m	$\varepsilon_{n,0}$	$n^{5/2}\varepsilon_{n,0}$	$n\,\varepsilon_{n,3/2}$
100	10	0.0000134495	1.345	1.744
200	15	0.0000024057	1.361	1.752
400	20	0.0000004278	1.369	1.756
800	29	0.0000000758	1.373	1.757
1000	32	0.0000000434	1.374	1.757
2000	45	0.0000000076	1.376	1.752
4000	64	0.0000000013	1.377	1.731
6000	78	0.0000000004	1.376	1.694
8000	90	0.0000000002	1.373	1.640
10000	100	0.0000000001	1.366	1.569

11 The Galerkin method

In this section, we are going to consider the Galerkin method (2.4), which is often also called the finite section method. We restrict ourselves to the consideration of the case $v = \varphi$ in the non-weighted space $\mathbf{L}^2(-1,1)$, that is, $\sigma \equiv 1$. (In what follows we will omit the symbol σ.) The results concerning this case and presented in the following were obtained in the paper [28], where beside the system $\{\widetilde{u}_n\}_{n=0}^{\infty}$ three similar systems of ansatz functions were considered. An interesting feature of the approach developed there is the fact that the local principle in Theorem 3.5 is not only employed to solve an invertibility problem connected with the stability of an approximation method (in fact, we do not have to do this here, since we can reduce the stability of (2.4) to the stability of a related Galerkin method in $\mathbf{L}^2(\mathbb{T})$ for which the local theory based on Theorem 3.5 was already developed in [45]), but we use the second part of the local principle to show an identity that allows the transformation of our problem to the space $\mathbf{L}^2(\mathbb{T})$.

We only mention here that for the case $\sigma = \varphi^{-1}$ the Galerkin method was considered in [33], where the transformation to $\mathbf{L}^2(\mathbb{T})$ is somewhat easier (due to the mapping properties (2.2) of the operator S with respect to $\widetilde{u}_n$) and does not require the application of the local principle we have just mentioned.

We remark that our approach allows in principle to derive stability conditions for (2.4), where A is an arbitrary operator from $\mathrm{alg}(S, \mathbf{PC}[-1,1])$, that is, the smallest closed subalgebra of $\mathcal{L}(\mathbf{L}^2)$ that contains S and all operators of multiplication by piecewise continuous functions. We give the conditions in explicit form for the operators $A = aI + bS$ and $A = aI + SbI$, where $a, b \in \mathbf{PC}$.

11.1 Preliminary results

In the sequel, we will often make use of isometric isomorphisms between certain spaces in order to reduce the stability problem under investigation to an equivalent one in a different space. Evidently, the mapping J defined by

$$J \sum_{n=0}^{\infty} \xi_n \widetilde{u}_n = \sum_{n=0}^{\infty} \xi_n e_n$$

is an isometric isomorphism between $\mathbf{L}^2$ and the Hardy space $\mathbf{H}^2(\mathbb{T}) = \mathrm{im}\, P_{\mathbb{T}} \subset \mathbf{L}^2(\mathbb{T})$. Evidently, by the isomorphism J_M,

$$J_M \sum_{n=0}^{\infty} \xi_n e_n = \{\xi_n\}_{n=0}^{\infty},$$

the space $\mathbf{H}^2(\mathbb{T})$ can in turn be identified with the sequence space

$$l^2 = \left\{ \{\xi_n\}_{n=0}^{\infty} : \xi_n \in \mathbb{C} \quad \text{and} \quad \|\{\xi_n\}\|_{l^2}^2 := \sum_{n=0}^{\infty} |\xi_n|^2 < \infty \right\}.$$

We remark that an operator $B \in \mathcal{L}(l^2)$ is associated with an infinite matrix $(b_{jk})_{j,k=0}^{\infty}$ such that for $\xi = \{\xi_n\} \in l^2$ the j-th component of $B\xi$ equals

$$\sum_{k=0}^{\infty} b_{jk}\xi_k, \qquad j = 0, 1, \ldots .$$

Evidently, we have $b_{jk} = \langle BJ_M e_k, J_M e_j\rangle_{l^2} = \langle J_M^{-1} BJ_M e_k, e_j\rangle_{\mathbf{L}^2(\mathbb{T})}$. Thus, via the isomorphism J_M an operator $A \in \mathcal{L}(\mathbf{H}^2(\mathbb{T}))$ can always be identified with the infinite matrix

$$A_M := J_M A J_M^{-1} = \left(\langle Ae_k, e_j\rangle_{\mathbf{L}^2(\mathbb{T})}\right)_{j,k=0}^{\infty} = \left(\langle J^{-1}AJ\widetilde{u}_k, \widetilde{u}_j\rangle_{\mathbf{L}^2(-1,1)}\right)_{j,k=0}^{\infty}.$$

The flip operator $C \in \mathcal{L}(\mathbf{L}^2(\mathbb{T}))$ is defined by $(Cu)(t) := t^{-1}u(t^{-1})$. Let $a \in \mathbf{L}^\infty(\mathbb{T})$. Now we can introduce the Toeplitz operator $T(a)$ and the Hankel operator $H(a)$ generated by a :

$$T(a) := P_{\mathbb{T}} a P_{\mathbb{T}} \,|\, _{\mathbf{H}^2(\mathbb{T})}, \qquad H(a) := P_{\mathbb{T}} a C P_{\mathbb{T}} \,|\, _{\mathbf{H}^2(\mathbb{T})}.$$

This definition implies the well-known fact that Hankel operators generated by continuous functions are compact. Furthermore, it is well-known that these operators have the matrix representations

$$(T(a))_M = (a_{j-k})_{j,k=0}^{\infty}, \qquad (H(a))_M = (a_{j+k+1})_{j,k=0}^{\infty},$$

where a_i denotes the i-th Fourier coefficient of a with respect to the system $\{e_k\}_{k=-\infty}^{\infty}$.

For a function $a \in \mathbf{L}^\infty(\mathbb{T})$ we define

$$\widetilde{a}(t) := a(t^{-1}).$$

With this notation we can add two very important relations for Toeplitz and Hankel operators of products:

$$T(ab) = T(a)T(b) + H(a)H(\widetilde{b}), \qquad H(ab) = H(a)T(\widetilde{b}) + T(a)H(b) \tag{11.1}$$

(see e. g. [6] for the assertions on Toeplitz and Hankel operators).

We will reduce the investigation of the stability of our finite section method (2.4) in $\mathbf{L}^2$ to a related problem in $\mathbf{L}^2(\mathbb{T})$, for which the results to be presented in the rest of this subsection were achieved in [45]. We consider operators on $\mathbf{L}^2(\mathbb{T})$. Let $\mathcal{H}$ be the Banach algebra of all bounded sequences $\{B_n\}_{n=1}^{\infty}$, $B_n \in \mathcal{L}(\mathbf{L}^2(\mathbb{T}))$, equipped with component-wise operations and the supremum norm. By $\mathcal{D}$ we denote the smallest closed subalgebra of $\mathcal{H}$ that contains the constant sequences $\{P_{\mathbb{T}}\}$, $\{C\}$ and $\{aI\}$ for $a \in \mathbf{PC}(\mathbb{T})$ and the sequences $\{P_{ln}^{\mathbb{T}}\}_{n=1}^{\infty}$ for every positive integer l. For $B \in \mathcal{L}(\mathbf{L}^2(\mathbb{T}))$ we consider the finite section method with the approximate operators

$$B_n := P_n^{\mathbb{T}} B P_n^{\mathbb{T}} + Q_n^{\mathbb{T}}, \tag{11.2}$$

where $Q_n^{\mathbb{T}} = I - P_n^{\mathbb{T}}$. We further define the operator $(W_{\mathbb{R}} f)(t) := f(-t)$ for $f \in \mathbf{L}^2(\mathbb{R})$, and by $\chi_{\alpha,\beta}$ we denote the characteristic function of the (bounded or unbounded) interval (α, β). Now we introduce two families of homomorphisms W_t $(t \in \mathbb{T},\ \mathrm{Im}\, t \geq 0)$, where

$$W_t : \mathcal{D} \to \mathcal{L}(\mathbf{L}^2(\mathbb{R})), \qquad t = \pm 1,$$

$$W_t : \mathcal{D} \to \mathcal{L}(\mathbf{L}_2^2(\mathbb{R})), \qquad \mathrm{Im}\, t > 0 \qquad (\mathbf{L}_2^2(\mathbb{R}) := \mathbf{L}^2(\mathbb{R}) \times \mathbf{L}^2(\mathbb{R})),$$

and W^l $(l \in \{0, 1, 2 \ldots\})$, where

$$W^l : \mathcal{D} \to \mathcal{L}(\mathbf{L}^2(\mathbb{T})), \qquad l = 0,$$

$$W^l : \mathcal{D} \to \mathcal{L}(\mathbf{L}_2^2(\mathbb{T})), \qquad l > 0 \qquad (\mathbf{L}_2^2(\mathbb{T}) := \mathbf{L}^2(\mathbb{T}) \times \mathbf{L}^2(\mathbb{T})).$$

The images of the generating elements of $\mathcal{D}$ are defined as follows (the one–sided limits $a(t+0)$, $a(t-0)$ etc. are to be understood according to the positive orientation of the unit circle):

$$W_t\{P_{kn}^{\mathbf{T}}\} = \begin{cases} \chi_{-k,k} I & , \quad t = \pm 1 \,, \\ \begin{pmatrix} \chi_{-k,k} I & 0 \\ 0 & \chi_{-k,k} I \end{pmatrix} & , \quad \operatorname{Im} t > 0 \,, \end{cases}$$

$$W^l\{P_{kn}^{\mathbf{T}}\} = \begin{cases} I & , \quad l = 0 \,, \\ \begin{pmatrix} I & 0 \\ 0 & I \end{pmatrix} & , \quad 0 < l < k \,, \\ \begin{pmatrix} Q_{\mathbf{T}} & 0 \\ 0 & Q_{\mathbf{T}} \end{pmatrix} & , \quad l = k \,, \\ \begin{pmatrix} 0 & 0 \\ 0 & 0 \end{pmatrix} & , \quad l > k \,, \end{cases}$$

$$W_t\{P_{\mathbf{T}}\} = \begin{cases} \chi_{0,\infty} I & , \quad t = \pm 1 \,, \\ \begin{pmatrix} \chi_{0,\infty} I & 0 \\ 0 & \chi_{-\infty,0} I \end{pmatrix} & , \quad \operatorname{Im} t > 0 \,, \end{cases}$$

$$W^l\{P_{\mathbf{T}}\} = \begin{cases} P_{\mathbf{T}} & , \quad l = 0 \,, \\ \begin{pmatrix} I & 0 \\ 0 & 0 \end{pmatrix} & , \quad l > 0 \,, \end{cases}$$

$$W_t\{aI\} = \begin{cases} a(t+0)Q_{\mathbb{R}} + a(t-0)P_{\mathbb{R}} \,, & t = \pm 1 \,, \\ \begin{pmatrix} a(t+0)Q_{\mathbb{R}} + a(t-0)P_{\mathbb{R}} & 0 \\ 0 & \widetilde{a}(t+0)Q_{\mathbb{R}} + \widetilde{a}(t-0)P_{\mathbb{R}} \end{pmatrix} , & \operatorname{Im} t > 0 \,, \end{cases}$$

$$W^l\{aI\} = \begin{cases} aI & , \; l = 0\,, \\ \begin{pmatrix} aI & 0 \\ 0 & \widetilde{a}I \end{pmatrix} & , \; l > 0\,, \end{cases}$$

$$W_t\{C\} = \begin{cases} \pm W_{\mathbb{R}} & , \; t = \pm 1\,, \\ \begin{pmatrix} 0 & I \\ I & 0 \end{pmatrix} & , \; \operatorname{Im} t > 0\,, \end{cases}$$

$$W^l\{C\} = \begin{cases} C & , \; l = 0\,, \\ \begin{pmatrix} 0 & I \\ I & 0 \end{pmatrix} & , \; l > 0\,. \end{cases}$$

Proposition 11.1 (see [45] or [21], Chapters 4,6)

(i) *The mappings W_t and W^l can be extended to continuous $*$-homomorphisms of the algebra $\mathcal{D}$ into $\mathcal{L}(L^2(\mathbb{R}))$ or $\mathcal{L}(L^2_2(\mathbb{R}))$ and $\mathcal{L}(L^2(\mathbb{T}))$ or $\mathcal{L}(L^2_2(\mathbb{T}))$, respectively.*

(ii) *A sequence $\{B_n\} \in \mathcal{D}$ is stable if and only if all operators $W_t\{\widetilde{B}_n\}$, $W^l\{\widetilde{B}_n\}$ for $t \in \mathbb{T}$, $\operatorname{Im} t \geq 0$, and $l \in \{0, 1, 2, \ldots\}$ are invertible.*

Remark 11.2 *Since in contrast to the scheme considered in Section 2 the approximation operators B_n are defined on the whole space $\mathbf{L}^2(\mathbb{T})$ rather than on finite-dimensional subspaces, the notion of stability here means that $B_n \in \mathcal{GL}(\mathbf{L}^2(\mathbb{T}))$ for all n greater than some n_0 and $\sup\limits_{n>n_0} \left\|B_n^{-1}\right\|_{\mathcal{L}(\mathbf{L}^2(\mathbb{T}))} < \infty$.*

11.2 Transformation of the operators

In this subsection we are going to compute the transformations JAJ^{-1} of the operators $A \in \mathcal{L}(\mathbf{L}^2)$, for which we consider the Galerkin method (2.4), to the Hardy space $\mathbf{H}^2(\mathbb{T})$, or sometimes more precisely their matrix representations $(JAJ^{-1})_M \in \mathcal{L}(l^2)$. First let us note that Lemma 4.20 immediately yields the representation

$$JVJ^{-1} = T(e_1), \qquad JV^*J^{-1} = T(e_{-1}). \tag{11.3}$$

We agree upon the following notation: If the difference of two linear bounded operators A and B on a Banach space is a compact operator, we will briefly write $A = B + comp$. Now we are able to consider the transformation of multiplication operators.

Lemma 11.3 *Let $a \in \mathbf{C}[-1,1]$. Then the operator aI of multiplication by a has the representation*

$$JaJ^{-1} = T(\widehat{a}) + comp.$$

Proof. First we show the assertion for $e^n(x) = x^n$ by induction. Because of (11.3) and the equation $eI = \frac{1}{2}(V + V^*)$, the assertion is obvious for $n = 0$ and $n = 1$. If we assume it to be true for some $n \geq 1$, we have due to (11.1)

$$Je^{n+1}J^{-1} = [T(\widehat{e^n}) + comp.][T(\widehat{e}) + comp.] = T(\widehat{e^{n+1}}) + comp.$$

Thus, the assertion is true for polynomials. Since $\|T(\widehat{a})\|_{\mathcal{L}(\mathbf{H}^2(\mathbf{T}))} \leq \|a\|_\infty$, the general case follows from Weierstraß' approximation theorem. ■

Lemma 11.4 *Let $-1 \leq c < d \leq 1$ and $\chi := \chi_{c,d}$. Then for the multiplication operator χI we have the relation*

$$J\chi J^{-1} = T(\widehat{\chi}) - H(e_{-1}\widehat{\chi}).$$

Proof. We put $A := \arccos c, B := \arccos d \in [-\pi, 0]$. First, a simple computation yields

$$\langle \widehat{\chi}, e_n \rangle_{\mathbf{L}^2(\mathbf{T})} = \frac{1}{2\pi} \int_{-\pi}^{\pi} \widehat{\chi}(e^{is})\, e^{-ins}\, \mathrm{d}s = \frac{1}{\pi} \int_A^B \cos ns\, \mathrm{d}s.$$

Using the trigonometric representation of U_n, we obtain

$$\langle \chi \widetilde{u}_k, \widetilde{u}_j \rangle_{\mathbf{L}^2(-1,1)} = \frac{1}{\pi} \int_A^B [\cos(j-k)s - \cos(j+k+2)s]\, \mathrm{d}s, \quad j,k = 0,1,2,\ldots. \quad ■$$

Combining the last two lemmas, we get a result for operators of multiplication by arbitrary piecewise continuous functions.

Lemma 11.5 *For every $a \in \mathbf{PC}$ we have*

$$JaJ^{-1} = T(\widehat{a}) - H(e_{-1}\widehat{a}) + comp.$$

Proof. First we consider a function a with a finite number of jumps $t_1 < t_2 < \ldots < t_n$ and put $t_0 := -1$, $t_{n+1} := 1$. Obviously, there exist continuous functions f_j ($j = 0, \ldots, n$) such that $a = \sum_{j=0}^{n} \chi_{t_j, t_{j+1}} f_j$. It is sufficient to consider one summand of this kind (we omit the indices). According to (11.1) and Lemmas 11.3 and 11.4, we have

$$J\chi f J^{-1} = [T(\widehat{\chi}) - H(e_{-1}\widehat{\chi})][T(\widehat{f}) + comp.]$$

$$= T(\widehat{\chi f}) - H(e_{-1}\widehat{\chi f}) + comp.$$

By approximation in $\mathbf{L}^\infty$, we get the assertion for arbitrary $a \in \mathbf{PC}$. ■

Our next concern is the representation of the singular integral operator S. For this end, we need several auxiliary considerations. It is well-known (see [20], Lemma I.4.2) that the operator $w^{-1}SwI$ belongs to $\mathcal{L}(L^2)$, where $w := \varphi^{\frac{1}{2}}$.

Lemma 11.6 *We have the relation*

$$Jw^{-1}SwJ^{-1} = -T(\psi) - H(e_{-1}\psi).$$

Proof. First we note that

$$\langle \psi, e_n \rangle_{\mathbf{L}^2(\mathbf{T})} = \frac{1-(-1)^n}{\pi n i}, \quad n \neq 0, \qquad \langle \psi, e_0 \rangle_{\mathbf{L}^2(\mathbf{T})} = 0.$$

Now we put $(Jw^{-1}SwJ^{-1})_M = (a_{jk})_{j,k=0}^{\infty}$ and compute a_{jk}. We have

$$a_{jk} = \langle w^{-1}Sw\widetilde{u}_k, \widetilde{u}_j \rangle_{\mathbf{L}^2} = i \int_{-1}^{1} T_{k+1}(x) U_j(x)\, \mathrm{d}x$$

$$= -\frac{2i}{\pi} \int_{-\pi}^{0} \cos(k+1)s \sin(j+1)s\, \mathrm{d}s = -\frac{1-(-1)^{j-k}}{(j-k)\pi i} - \frac{1-(-1)^{j+k+2}}{(j+k+2)\pi i}$$

if $k \neq j$ and $a_{jk} = -\dfrac{1-(-1)^{j+k+2}}{(j+k+2)\pi i}$ if $k = j$. Thus, the proof is complete. ∎

Since $wSw^{-1}I = (w^{-1}SwI)^*$ (cf. [20], Chapt. I, §7), Lemma 11.6 implies that

$$\frac{1}{2} J(w^{-1}SwI + wSw^{-1}I)J^{-1} = -T(\psi) + comp. \tag{11.4}$$

Now we need some results on singular integral operators on the half-line which can be found in [46, Chapter I]. Let $\mathbb{R}^+ = (0, \infty)$, $\mathbb{R}^- = (-\infty, 0)$, and let $\mathrm{alg}(S_{\mathbb{R}^+})$ denote the smallest closed subalgebra of $\mathcal{L}(\mathbf{L}^2(\mathbb{R}^+))$ that contains $S_{\mathbb{R}^+}$ and the identity operator.

Lemma 11.7 (cf. [46], Chapter I or [21], 2.1.2) *The maximal ideal space of the commutative C^*-algebra* $\mathrm{alg}(S_{\mathbb{R}^+})$ *can be identified with the two-point compactification* $\overline{\mathbb{R}}$ *of* $\mathbb{R}$. *For $\gamma \in (-\frac{1}{2}, \frac{1}{2})$ the operator S_γ defined by*

$$(S_\gamma u)(x) = \frac{1}{\pi i} \int_0^{\infty} \left(\frac{t}{x}\right)^{\gamma} \frac{u(t)}{t-x}\, \mathrm{d}t.$$

is contained in $\mathrm{alg}(S_{\mathbb{R}^+})$, *and the Gelfand transform of S_γ is the mapping*

$$z \mapsto \coth(z + i(1/2 - \gamma))\pi, \qquad z \in \overline{\mathbb{R}}.$$

It is well-known (see [15, No. 407]) that for $t \in [-1, 1]$ we have the identity

$$(1-t)^{1/2} = \sum_{k=0}^{\infty} (-1)^k \binom{1/2}{k} t^k,$$

where the series is absolutely convergent. The substitution

$$t = \left(\frac{2 \tanh \pi z}{1 + \tanh^2 \pi z} \right)^2$$

and a little computation yield

$$\tanh \pi z = \sum_{k=1}^{\infty} (-1)^{k-1} \binom{1/2}{k} \left(\frac{2 \tanh \pi z}{1 + \tanh^2 \pi z} \right)^{2k-1} \tag{11.5}$$

for $z \in \mathbb{R}$. Using this result and some relations between hyperbolic functions, Lemma 11.7 results in

$$S_{\mathbb{R}^+} = \sum_{k=1}^{\infty}(-1)^{k-1}\binom{1/2}{k}\left(\frac{S_{1/4}+S_{-1/4}}{2}\right)^{2k-1}. \tag{11.6}$$

In the following, we will use the notation B^π for the coset $B + \mathcal{K}(\mathbf{L}^2(\mathbb{R}))$ if $B \in \mathcal{L}(\mathbf{L}^2(\mathbb{R}))$. Furthermore, we will briefly write χ for $\chi_{-1,1}$. Now we identify $S \in \mathcal{L}(\mathbf{L}^2(-1,1))$ with $\chi S_{\mathbb{R}}\chi I$, and we are going to prove a representation for JSJ^{-1} up to a compact summand using local techniques on the real line.

Put $\mathcal{B}_0 = \operatorname{alg}(S_{\mathbb{R}}, \mathbf{PC}(\dot{\mathbb{R}}))$, that is, $\mathcal{B}_0$ is the smallest closed subalgebra of $\mathcal{L}(\mathbf{L}^2(\mathbb{R}))$ containing $S_{\mathbb{R}}$ and the operators of multiplication by functions that are piecewise continuous on the one-point compactification $\dot{\mathbb{R}}$ of $\mathbb{R}$. This algebra contains $\mathcal{K}(\mathbf{L}^2(\mathbb{R}))$ (see [19, Lemma 4.1] or [46, Prop. 3.4]), and so we can form the quotient algebra $\mathcal{B} := \mathcal{B}_0/\mathcal{K}(\mathbf{L}^2(\mathbb{R}))$. Let $\mathcal{C}$ be the subalgebra of all cosets $(fI)^\pi$ such that $f \in \mathbf{C}(\dot{\mathbb{R}})$. Evidently, $\mathcal{C}$ is contained in the center of $\mathcal{B}$ (see [20, Th. I.4.3]). One can easily show that $\mathcal{C}$ is isometrically isomorphic to $\mathbf{C}(\dot{\mathbb{R}})$. Hence, the maximal ideal space $M(\mathcal{C})$ is homeomorphic to $\dot{\mathbb{R}}$, and the Gelfand transform of $(fI)^\pi \in \mathcal{C}$ is f itself. For $\tau \in \dot{\mathbb{R}}$ define the local ideals $J_\tau \subset \mathcal{B}$ according to Theorem 3.5. Further, let $M_{\tau_0} = \{(fI)^\pi : f \in m_{\tau_0}\}$, where

$$m_{\tau_0} = \{f \in \mathbf{C}(\dot{\mathbb{R}}) : 0 \le f(\tau) \le 1, f(\tau) \equiv 1 \text{ in some neighbourhood of } \tau_0\}.$$

Then Remark 3.7 says that $J_\tau = \{x \in \mathcal{B} : x \overset{M_\tau}{\sim} 0\}$. Using this remark, we are going to show that

$$[\chi S_{\mathbb{R}}\chi I]^\pi - \sum_{k=1}^{\infty}(-1)^{k-1}\binom{1/2}{k}\left[\left(\frac{w^{-1}\chi S_{\mathbb{R}}\chi wI + w\chi S_{\mathbb{R}}\chi w^{-1}I}{2}\right)^{2k-1}\right]^\pi \in J_\tau \tag{11.7}$$

for all $\tau \in \dot{\mathbb{R}}$, which will imply the assertion of the following lemma.

Lemma 11.8 *We have the relation*

$$[\chi S_{\mathbb{R}}\chi I]^\pi = \sum_{k=1}^{\infty}(-1)^{k-1}\binom{1/2}{k}\left[\left(\frac{w^{-1}\chi S_{\mathbb{R}}\chi wI + w\chi S_{\mathbb{R}}\chi w^{-1}I}{2}\right)^{2k-1}\right]^\pi.$$

Proof. We will show (11.7). Then the assertion follows with the help of Theorem 3.5 and Remark 3.6.

For $\tau \in \mathbb{R}\setminus[-1,1]$ it is evident that both sides of the equation in the lemma are M_τ-equivalent to 0 and hence in J_τ. Now let $\tau \in (-1,1)$ be fixed, $f \in m_\tau$, $\operatorname{supp} f \subset (-1,1)$. Then we have

$$S_{\mathbb{R}}^\pi - [\chi S_{\mathbb{R}}\chi I]^\pi \in J_\tau, \tag{11.8}$$

because

$$(fI)^\pi(\chi S_{\mathbb{R}}\chi I)^\pi - (fI)^\pi S_{\mathbb{R}}^\pi = \left(f^{1/2}\chi S_{\mathbb{R}}\chi f^{1/2}I\right)^\pi - \left(f^{1/2}S_{\mathbb{R}}f^{1/2}I\right)^\pi = 0.$$

Furthermore,

$$(fI)^\pi\left[S_{\mathbb{R}} - \frac{w^{-1}\chi S_{\mathbb{R}}\chi wI + w\chi S_{\mathbb{R}}\chi w^{-1}I}{2}\right]^\pi = 0. \tag{11.9}$$

To explain the latter relation, we introduce a positive continuous function $\widetilde{w}$ on $\dot{\mathbb{R}}$ such that $\widetilde{w}$ coincides with w on supp f. Then we have

$$(fw^{-1}\chi S_{\mathbb{R}}\chi wI)^\pi = (f^{1/2}w^{-1}\chi S_{\mathbb{R}}\chi w f^{1/2}I)^\pi$$
$$= (f^{1/2}\widetilde{w}^{-1}\chi S_{\mathbb{R}}\chi\widetilde{w}f^{1/2}I)^\pi = (\chi S_{\mathbb{R}}\chi\widetilde{w}^{-1}\widetilde{w}fI)^\pi = (f\chi S_{\mathbb{R}}\chi I)^\pi.$$

Since J_τ is a closed two-sided ideal, we get from (11.9)

$$\sum_{k=1}^{\infty}(-1)^{k-1}\binom{1/2}{k}[S_{\mathbb{R}}^{2k-1}]^\pi$$
$$-\sum_{k=1}^{\infty}(-1)^{k-1}\binom{1/2}{k}\left[\left(\frac{w^{-1}\chi S_{\mathbb{R}}\chi wI + w\chi S_{\mathbb{R}}\chi w^{-1}I}{2}\right)^{2k-1}\right]^\pi \in J_\tau$$

for $\tau \in (-1,1)$. (Note that the fact that J_τ is an ideal implies that for $x-y \in J_\tau$ we also have $x^m - y^m \in J_\tau$ for every positive integer m.) The second series converges because of (11.4) and the relation $\|T(\psi)\|_{\mathcal{L}(\mathbf{H}^2(\mathbf{T}))} = \|\psi\|_\infty = 1$. Since $S_{\mathbb{R}}^{2k-1} = S_{\mathbb{R}}$, the first one equals $S_{\mathbb{R}}^\pi$. Together with (11.8), this shows that (11.7) holds for all $\tau \in (-1,1)$.

Now we consider the case $\tau = -1$. We assume $f \in m_{-1}$, supp $f \subset (-\infty, 1)$, and we define $v(x) := (1+x)^{1/4}$. Then it is easy to see that

$$(fI)^\pi\left[\frac{v^{-1}\chi_{-1,\infty}S_{\mathbb{R}}\chi_{-1,\infty}vI + v\chi_{-1,\infty}S_{\mathbb{R}}\chi_{-1,\infty}v^{-1}I}{2}\right]^\pi$$
$$-(fI)^\pi\left[\frac{w^{-1}\chi S_{\mathbb{R}}\chi wI + w\chi S_{\mathbb{R}}\chi w^{-1}I}{2}\right]^\pi = 0.$$

(Here the factor $(1-x)^{1/4}$ of w is treated in the same way as w itself in (11.9).) Thus, we get

$$\sum_{k=1}^{\infty}(-1)^{k-1}\binom{1/2}{k}\left[\left(\frac{w^{-1}\chi S_{\mathbb{R}}\chi wI + w\chi S_{\mathbb{R}}\chi w^{-1}I}{2}\right)^{2k-1}\right]^\pi \tag{11.10}$$
$$-\sum_{k=1}^{\infty}(-1)^{k-1}\binom{1/2}{k}\left[\left(\frac{\chi_{-1,\infty}v^{-1}S_{\mathbb{R}}v\chi_{-1,\infty}I + \chi_{-1,\infty}vS_{\mathbb{R}}v^{-1}\chi_{-1,\infty}I}{2}\right)^{2k-1}\right]^\pi \in J_{-1}.$$

The convergence of the second series of (11.10) can be obtained as follows: Let $(Mu)(x) := u(x+1)$ and consider the isometric isomorphism $A \mapsto MAM^{-1}$ between $\mathcal{L}(\mathbf{L}^2(\mathbb{R}^+))$ and $\mathcal{L}(\mathbf{L}^2(-1,\infty))$, which transforms $S_{\mathbb{R}^+}$ into $S_{(-1,\infty)}$ and $S_{\pm 1/4}$ into $v^{-1}S_{(-1,\infty)}vI$ and $vS_{(-1,\infty)}v^{-1}I$, respectively. Thus, the second series in (11.10) converges to $S_{(-1,\infty)}^\pi$ because of (11.6). Since $S_{(-1,\infty)}^\pi - [\chi S_{\mathbb{R}}\chi I]^\pi \in J_{-1}$, we have

$$[\chi S_{\mathbb{R}}\chi I]^\pi - \sum_{k=1}^{\infty}(-1)^{k-1}\binom{1/2}{k}\left[\left(\frac{w^{-1}\chi S_{\mathbb{R}}\chi wI + w\chi S_{\mathbb{R}}\chi w^{-1}I}{2}\right)^{2k-1}\right]^\pi \in J_{-1}.$$

The case $\tau = 1$ is treated analogously. ■

Corollary 11.9 *Lemma 11.8 together with equation (11.4) now shows the representation*

$$JSJ^{-1} = -\sum_{k=1}^{\infty}(-1)^{k-1}\binom{1/2}{k}(T(\psi))^{2k-1} + comp.,$$

where the series converges uniformly, since $\|T(\psi)\|_{\mathcal{L}(\mathbf{H}^2(\mathbb{T}))} = 1$.

Combining this relation with Lemma 11.5, we can summarize our results as follows:

Proposition 11.10 *For $a, b \in \mathbf{PC}$, we have the following transformation of the operator $aI + bS$:*

$$J(aI+bS)J^{-1} = T(\widehat{a}) - H(e_{-1}\widehat{a})$$

$$- \left[T(\widehat{b}) - H(e_{-1}\widehat{b})\right]\sum_{k=1}^{\infty}(-1)^{k-1}\binom{1/2}{k}(T(\psi))^{2k-1} + comp.$$

An analogous transformation holds for the operator $aI + SbI$.

11.3 Main result

Now we will apply the results of Proposition 11.1 to our problem. The applicability of these results is based on the following considerations: Let $A \in \mathcal{L}(\mathbf{L}^2)$. Then the stability of the approximate operator sequence $A_n := A_{n,P}$ used in the Galerkin method (2.4) is equivalent to the stability of (11.2) for the operator $B := JAJ^{-1}P_{\mathbb{T}} + Q_{\mathbb{T}} \in \mathcal{L}(\mathbf{L}^2(\mathbb{T}))$. Indeed,

$$B_n = P_n^{\mathbb{T}}(JAJ^{-1}P_{\mathbb{T}} + Q_{\mathbb{T}})P_n^{\mathbb{T}} + Q_n^{\mathbb{T}} = JA_nJ^{-1}P_{\mathbb{T}} + Q_{\mathbb{T}}P_n^{\mathbb{T}} + Q_n^{\mathbb{T}}$$

is invertible in $\mathbf{L}^2(\mathbb{T})$ if and only if A_n has an inverse A_n^{-1} in $\operatorname{im} P_n$, in which case

$$B_n^{-1} = JA_n^{-1}P_nJ^{-1}P_{\mathbb{T}} + Q_{\mathbb{T}}P_n^{\mathbb{T}} + Q_n^{\mathbb{T}},$$

which also yields the equivalence of the uniform boundedness of the inverses. In what follows we compute the images of the sequence $\{B_n\}$ under the homomorphisms W_t and W^l and discuss their invertibility. Here we make use of the fact that $\{K\}$ is contained in $\mathcal{D}$ and belongs to the kernel of W_t ($\operatorname{Im} t \geq 0$) and of W^l ($l \geq 1$) for every compact operator K (see [45]). Thus, we can neglect the compact summand in the representation given in Proposition 11.10. We restrict ourselves to the investigation of the operator $A = aI + bS$ ($a, b \in \mathbf{PC}$). We only mention that one can obtain the same stability conditions in the case of the operator $aI + SbI$. For the sake of brevity, we will use the notation $c := \widehat{a} - \widehat{b\psi}$.

- Evidently, $W^0\{B_n\} = B$, that is, we have the condition that $B = JAJ^{-1}P_{\mathbb{T}} + Q_{\mathbb{T}}$ is invertible in $\mathbf{L}^2(\mathbb{T})$ and, hence, A is invertible in $\mathbf{L}^2$.

- A little thought (cf. also the series computations in the previous subsection) yields

$$W^1\{B_n\} = \begin{pmatrix} Q_{\mathbf{T}} c Q_{\mathbf{T}} + P_{\mathbf{T}} & 0 \\ 0 & I \end{pmatrix},$$

and this operator is invertible if and only if $Q_{\mathbf{T}} c Q_{\mathbf{T}} + P_{\mathbf{T}}$ is invertible. Since $Q_{\mathbf{T}} c Q_{\mathbf{T}} + P_{\mathbf{T}} = (cQ_{\mathbf{T}} + P_{\mathbf{T}})(I + P_{\mathbf{T}} c Q_{\mathbf{T}})^{-1}$, the invertibility of $W^1\{B_n\}$ is equivalent to the invertibility of $cQ_{\mathbf{T}} + P_{\mathbf{T}}$, or, which is again equivalent, of $cP_{\mathbf{T}} + Q_{\mathbf{T}}$ in $\mathbf{L}^2(\mathbb{T})$.

- For $l > 1$, the homomorphisms W^l always produce the identity operator in $\mathbf{L}^2_2(\mathbb{T})$.

- For $t = \pm 1$, we get

$$\begin{aligned} W_t\{B_n\} &= a(\pm 1)\chi_{0,1} I \\ &\pm b(\pm 1)\chi_{0,1} \sum_{k=1}^{\infty} (-1)^{k-1} \binom{1/2}{k} [\chi_{0,\infty} S_{\mathbb{R}} \chi_{0,\infty}]^{2k-1} \chi_{0,1} I + \chi_{\mathbb{R}\setminus(0,1)} I. \end{aligned}$$

It will prove advantageous to investigate the restriction of these operators to $\mathbf{L}^2(\mathbb{R}^+)$.

Lemma 11.11 *If $W^1\{B_n\}$ is invertible, then so are $W_t\{B_n\}$ for $t = \pm 1$.*

Proof. Obviously, $\chi_{0,\infty} S_{\mathbb{R}} \chi_{0,\infty} I$ can be identified with $S_{\mathbb{R}^+}$. Using equation (11.5) and the identity $\tanh 2x = (2\tanh x)/(1 + \tanh^2 x)$, we obtain

$$\sum_{k=1}^{\infty} (-1)^{k-1} \binom{1/2}{k} (\tanh \pi z)^{2k-1} = \tanh \frac{\pi z}{2}. \qquad (11.11)$$

Now we introduce the Mellin transformation $\mathcal{M} \in \mathcal{L}(\mathbf{L}^2(\mathbb{R}^+), \mathbf{L}^2(\mathbb{R}))$ defined by

$$(\mathcal{M}u)(\lambda) = \int_0^{\infty} t^{1/2-\lambda i}\, u(t)\, dt$$

and the Fourier-Plancherel transformation $\mathcal{F} \in \mathcal{L}(\mathbf{L}^2(\mathbb{R}))$,

$$(\mathcal{F}u)(\lambda) = \int_{-\infty}^{\infty} e^{i\lambda t}\, u(t)\, dt.$$

According to [46, Prop. 2.2] one has $S_{\mathbb{R}^+} = \mathcal{M}^{-1}\gamma\mathcal{M}$ with $\gamma(z) = \tanh \pi z$. From equation (11.11) we get

$$W_{\pm 1}\{B_n\} \big|_{\mathbf{L}^2(\mathbb{R}^+)} = a(\pm 1)\chi_{0,1} I \pm b(\pm 1)\chi_{0,1}\mathcal{M}^{-1}\delta\mathcal{M}\chi_{0,1} I + \chi_{1,\infty} I \qquad (11.12)$$

with $\delta(z) = \tanh(\pi z/2)$. We further define $E : \mathbf{L}^2(\mathbb{R}^+) \to \mathbf{L}^2(\mathbb{R})$ by

$$(Eu)(t) = e^{-t/2} u(e^{-t}).$$

One can easily verify that E is an isometric isomorphism, that $\mathcal{M} = \mathcal{F}E$, and that $E\chi_{0,1}E^{-1} = \chi_{\mathbb{R}^+} I$. Thus, instead of (11.12) we can consider

$$a(\pm 1)\chi_{\mathbb{R}^+} I \pm b(\pm 1)\chi_{\mathbb{R}^+}\mathcal{F}^{-1}\delta\mathcal{F}\chi_{\mathbb{R}^+} I + \chi_{\mathbb{R}^-} I$$

in $\mathbf{L}^2(\mathbb{R})$. The invertibility of this Wiener-Hopf operator is equivalent to the condition

$$a(\pm 1) + \mu b(\pm 1) \neq 0 \qquad \text{for all} \quad \mu \in [-1, 1]$$

(cf. [11, Prop. 4.2]). These conditions, however, are necessary for the invertibility of the operator $[\widehat{a} - \widehat{b\psi}] P_{\mathbf{T}} + Q_{\mathbf{T}}$ (see Proposition 5.14). ■

- For $a, b \in \mathbf{PC}$ and $x \in (-1,1)$ we introduce the notations

$$c^+(x) := a(x-0) - b(x-0), \qquad c^-(x) := a(x+0) - b(x+0),$$

$$\widetilde{c}^+(x) := a(x-0) + b(x-0), \qquad \widetilde{c}^-(x) := a(x+0) + b(x+0).$$

Then for $\operatorname{Im} t > 0$ a short computation leads to

$$W_t\{B_n\} = \begin{pmatrix} \chi_{0,1}(c^+Q_{\mathbb{R}} + c^-P_{\mathbb{R}})\chi_{0,1}I & -t^{-1}\chi_{0,1}(\widetilde{c}^+Q_{\mathbb{R}} + \widetilde{c}^-P_{\mathbb{R}})\chi_{-1,0}I \\ -t\chi_{-1,0}(c^+Q_{\mathbb{R}} + c^-P_{\mathbb{R}})\chi_{0,1}I & \chi_{-1,0}(\widetilde{c}^+Q_{\mathbb{R}} + \widetilde{c}^-P_{\mathbb{R}})\chi_{-1,0}I \end{pmatrix}$$

$$+ \begin{pmatrix} I - \chi_{0,1}I & 0 \\ 0 & I - \chi_{-1,0}I \end{pmatrix}.$$

where c^+, c^-, $\widetilde{c}^+$, $\widetilde{c}^-$ have to be taken at the point $x = \operatorname{Re} t$.

Lemma 11.12 *For $\operatorname{Im} t > 0$, $W_t\{B_n\}$ is invertible if and only if $c^+\widetilde{c}^+c^-\widetilde{c}^- \neq 0$ and the origin lies outside the triangle which is formed by the points $\frac{c^-}{c^+}$, $\frac{\widetilde{c}^-}{\widetilde{c}^+}$, and 1.*

Proof. We denote the first summand of $W_t\{B_n\}$ by D. Evidently, $\mathbf{L}^2_2(\mathbb{R})$ decomposes into the direct sum of $\mathbf{L}^2(0,1) \times \mathbf{L}^2(-1,0)$ and $\mathbf{L}^2(\mathbb{R} \setminus (0,1)) \times \mathbf{L}^2(\mathbb{R} \setminus (-1,0))$. The first of these two subspaces contains the image of D, while the second one is contained in the kernel of D. Thus, we can consider D as an operator on $\mathbf{L}^2(0,1) \times \mathbf{L}^2(-1,0)$, where $P_{\mathbb{R}}$ and $Q_{\mathbb{R}}$ can be replaced by P and Q, respectively. Obviously, $W_t\{B_n\}$ is invertible if and only if D is invertible in $\mathbf{L}^2(0,1) \times \mathbf{L}^2(-1,0)$. The latter space can be identified in a natural way with $\mathbf{L}^2(-1,1)$. After this identification D gets the form

$$\begin{aligned} D &= \chi_{0,1}(c^+Q + c^-P)\chi_{0,1}I - t^{-1}\chi_{0,1}(\widetilde{c}^+Q + \widetilde{c}^-P)\chi_{-1,0}I \\ &\quad - t\chi_{-1,0}(c^+Q + c^-P)\chi_{0,1}I + \chi_{-1,0}(\widetilde{c}^+Q + \widetilde{c}^-P)\chi_{-1,0}I \\ &= (\chi_{0,1} - t\chi_{-1,0})\left[Q(c^+\chi_{0,1} - t^{-1}\widetilde{c}^+\chi_{-1,0})I + P(c^-\chi_{0,1} - t^{-1}\widetilde{c}^-\chi_{-1,0})I\right]. \end{aligned}$$

Since $t \neq 0$, the invertibility of this operator is equivalent to the conditions in this lemma (cf. Proposition 5.14). ■

Now we are in a position to summarize our results. We point out again that the preceding considerations lead to the same conditions in the case of the operator $aI + SbI$.

Theorem 11.13 *For $a, b \in \mathbf{PC}$ and $A = aI + bS$ or $A = aI + SbI$, the operator sequence $\{P_nAP_n\}$ is stable in $\mathbf{L}^2(-1,1)$ if and only if the following conditions are satisfied:*

(i) A is invertible on $\mathbf{L}^2$,

(ii) $[\widehat{a} - \widehat{b}\psi]P_{\mathbb{T}} + Q_{\mathbb{T}}$ is invertible on $\mathbf{L}^2(\mathbb{T})$, and

(iii) for every $x \in (-1,1)$ *the origin lies outside the triangle which is formed by the points* $\dfrac{c^-(x)}{c^+(x)}, \dfrac{\tilde{c}^-(x)}{\tilde{c}^+(x)}$, *and 1.*

Remark 11.14 *It should be pointed out that neither the invertibility of* $cP_{\mathbb{T}} + Q_{\mathbb{T}}$ *in* $\mathbf{L}^2(\mathbb{T})$ *nor the "local" conditions of Lemma 11.12 follow from the invertibility of* A. *The case* $b \equiv 0$, $a(x) = \operatorname{sgn} x$ *may serve as an example.*

12 A forward look

We close by mentioning some problems connected with the collocation method that are desirable (and expected) to be solved but that the author has not yet been able to cope with in the present paper.

The most significant imperfection of this paper, in the opinion of the author, is the rather restrictive condition that we have to impose on the coefficient b, except in the special case $v = \varphi$, $\sigma = \varphi^{-1}$. Recently, however, P. Junghanns and G. Mastroianni succeeded in proving the strong convergence of the operators $\widetilde{A}_n$ to $\widetilde{A} = aI - bw_\sigma^{-1}Sw_\sigma I$ for arbitrary $b \in \mathbf{PC}$ and arbitrary σ satisfying (1.4) (except the cases $\alpha = \frac{1}{2}$ or $\beta = \frac{1}{2}$) in the case that $v = \varphi$. This result gives rise to the hope that also the local theory can be carried out in this general situation.

The two homomorphisms $\mathcal{W}_0, \mathcal{W}_1$ are in a certain sense not sufficient (compare [47]). Thus, it might be necessary to search for further "hidden" homomorphisms generating stability conditions that are automatically satisfied if A and $\widetilde{A}$ are invertible in the cases we considered so far ($\sigma = \varphi^{-1}$ or $b(\pm 1) = 0$, in the system case always $\underline{b}(\pm 1) = 0$) but might become essential in the general case and in particular in the system case.

It is desirable not to restrict the investigations to the special operators $aI + bS$, but to treat a wider class. P. Junghanns obtained some first results in the consideration of the stability problem for sequences in the subalgebra $\mathcal{B}$ of $\mathcal{A}$ generated by all sequences $\{\widetilde{L_n^\sigma}(aI + bS)P_n^\sigma\}$ with $a, b \in \mathbf{PC}$ and the ideal $\mathcal{I}$ in the special case $v = \varphi$, $\sigma = \varphi^{-1}$. A suitable approach to the invertibility modulo the ideal $\mathcal{I}$ in this situation is via the local principle of Allan in Theorem 3.5. If we choose the set $\mathcal{C} = \{\{\widetilde{L_n^\sigma fP_n^\sigma}\} + \mathcal{I} : f \in \mathbf{C}[-1,1]\}$ as a central subalgebra of $\mathcal{B}/\mathcal{I}$, the quotient algebras with respect to the local ideals J_τ can be described by the two-projection-theorem [44, Th. 1.16] for $\tau \in (-1, 1)$ and are generated by one element for $\tau = \pm 1$.

Furthermore, the conditions under which we are able to prove error estimates in Sobolev norms are very strong, but the numerical results we obtained show that such strong conditions are not really necessary. Thus, it is desirable to refine these results, in particular to weaken the conditions under which the boundedness and the invertibility of singular integral operators in Sobolev spaces can be proved.

Finally we hope that an analysis of the results obtained in [22] will allow to further reduce the computational complexity of the collocation method to $O(n \ln n)$ operations.

Index of symbols and notations

References

[1] B. A. Amosov. On the approximate solution of elliptic pseudodifferential equations on a smooth closed curve (Russian). *Z. Anal. Anw.*, 9:545–563, 1990.

[2] V. M. Badkov. Convergence in mean and almost everywhere of Fourier series in polynomials orthogonal on a segment (Russian). *Mat. Sb.*, Vol. 95(137), No. 2(10):229–262, 1974.

[3] Yu. M. Berezanski. *Expansions with Respect to Eigenfunctions of Selfadjoint Operators (Russian).* Naukova Dumka, Kiev, 1965.

[4] D. Berthold, W. Hoppe, and B. Silbermann. A fast algorithm for solving the generalized airfoil equation. *J. Comp. Appl. Math.*, 43:185–219, 1992.

[5] D. Berthold, W. Hoppe, and B. Silbermann. The numerical solution of the generalized airfoil equation. *Journal of Integral Equations and Applications*, 4(3):309–336, 1992.

[6] A. Böttcher and B. Silbermann. *Analysis of Toeplitz Operators.* Springer-Verlag, Berlin, Heidelberg, New York, 1990.

[7] M. Braun. *Algebraische Theorie und schnelle Invertierung von Löwner- und Pick-Matrizen.* Master's thesis, Technische Universität Karl-Marx-Stadt (now Chemnitz), 1988.

[8] M. R. Capobianco, P. Junghanns, U. Luther, and G. Mastroianni. Weighted uniform convergence of the quadrature method for Cauchy singular integral equations. In A. Böttcher and I. Gohberg, editors, *Singular Integral Operators and Related Topics. Operator Theory Advances and Applications, Vol. 90*, pages 153–181. Birkhäuser Verlag, 1996.

[9] Z. Ditzian and V. Totik. *Moduli of Smoothness.* Springer Verlag, Berlin, Heidelberg, New York, 1987.

[10] M. L. Dow and D. Elliott. The numerical solution of singular integral equations over $(-1,1)$. *SIAM J. Numer. Anal.*, 16(1):115–134, 1979.

[11] R. Duduchava. *Integral Equations with Fixed Singularities.* Teubner Verlag, Leipzig, 1979.

[12] D. Elliott. The classical collocation method for singular integral equations. *SIAM J. Numer. Anal.*, 19(4):816–832, 1982.

[13] D. Elliott. Orthogonal polynomials associated with singular integral equations having a Cauchy kernel. *SIAM J. Math. Anal*, 13(6):1041–1052, 1982.

[14] D. Elliott. A comprehensive approach to the approximate solution of singular integral equations over the arc $(-1,1)$. *Journal of Integral Equations and Applications*, 2:59–94, 1989.

[15] G. M. Fichtenholz. *Differential- und Integralrechnung*. Deutscher Verlag der Wissenschaften, Berlin, 1990.

[16] G. Freud. *Orthogonale Polynome*. Deutscher Verlag der Wissenschaften, Berlin, 1969.

[17] W. Gautschi. Computational aspects of orthogonal polynomials. In P. Nevai, editor, *Orthogonal Polynomials: Theory and Practice*, pages 181–216. Kluwer Academic Publishers, 1990.

[18] W. Gautschi and J. Wimp. Computing the Hilbert transform of a Jacobi weight function. *BIT*, 27:203–215, 1987.

[19] I. Gohberg and N. Krupnik. On the algebra generated by one-dimensional singular integral operators with piecewise continuous coefficients (Russian). *Funkts. Anal. Prilozhen.*, 4(3):26–36, 1970. English transl.: *Funct. Anal. Appl.* 4(3):193–201, 1970.

[20] I. Gohberg and N. Krupnik. *Einführung in die Theorie der eindimensionalen singulären Integraloperatoren*. Birkhäuser Verlag, Basel, Boston, Stuttgart, 1979.

[21] R. Hagen, S. Roch, and B. Silbermann. *Spectral Theory of Approximation Methods for Convolution Equations*. Birkhäuser Verlag, Basel, Boston, Berlin, 1994.

[22] G. Heinig and K. Rost. Hartley transform representations of inverses of real Toeplitz-plus-Hankel matrices. *to appear in Numerical Functional Analysis and Optimization*.

[23] G. Heinig and K. Rost. *Algebraic Methods for Toeplitz-like Matrices and Operators*. Birkhäuser Verlag, Basel, Boston, Berlin, 1984.

[24] P. Junghanns. Numerical Analysis of Newton projection methods for nonlinear singular integral equations. *J. Comp. Appl. Math.*, 55(2):145–163, 1994.

[25] P. Junghanns. Product integration for the generalized airfoil equation. In E. Schock, editor, *Beiträge zur Angewandten Analysis und Informatik*, pages 171–188. Shaker Verlag, Aachen, 1994.

[26] P. Junghanns. Numerical solution of a free surface seepage problem from nonlinear channels. *Appl. Anal.*, 63:87–110, 1996.

[27] P. Junghanns and K. Müller. A collocation method for nonlinear Cauchy singular integral equations. *to appear in J. Comp. Appl. Math.*

[28] P. Junghanns, S. Roch, and U. Weber. Finite sections for singular integral operators by weighted Chebyshev polynomials. *Integral Equations Operator Theory*, 21:319–333, 1995.

[29] P. Junghanns and B. Silbermann. *Numerical Analysis of the quadrature method for solving linear and nonlinear singular integral equations*. Wiss. Schriftenreihe TU Karl-Marx-Stadt (now Chemnitz), 1988.

[30] P. Junghanns and B. Silbermann. Theorie der Näherungsverfahren für singuläre Integralgleichungen auf Intervallen. *Math. Nachr.*, 103:199–244, 1981.

[31] P. Junghanns and B. Silbermann. Local theory of the collocation method for singular integral equations. *Integral Equations Operator Theory*, 7:791–807, 1984.

[32] P. Junghanns and U. Weber. *Local theory of a collocation method for Cauchy singular integral equations on an interval.* Preprint 97-10 TU Chemnitz, 1997.

[33] P. Junghanns and U. Weber. Local theory of projection methods for Cauchy singular integral equations on an interval. In M. Golberg, editor, *Boundary Integral Methods—Numerical and Mathematical Aspects*, pages 217–256. Comp. Mech. Publ., Boston, Southampton, 1999.

[34] A. I. Kalandiya. *Mathematical Methods of Two-dimensional Elasticity.* MIR, Moscow, 1975.

[35] Kozak. On the reduction method for multidimensional discrete convolutions (Russian). *Mat. Issled.*, 8(3):157–160, 1973.

[36] U. Luther. *Kollokations- und Quadraturformelverfahren für singuläre Integralgleichungen bzgl. modifizierter Systeme von Nullstellen orthogonaler Polynome.* Master's thesis, TU Chemnitz, 1995.

[37] U. Luther. *Generalized Besov Spaces and Cauchy Singular Integral Equations.* PhD thesis, TU Chemnitz, 1998.

[38] G. Mastroianni and M. G. Russo. Lagrange interpolation in weighted Besov spaces. *Constr. Approx.*, 15(2):257–289, 1999.

[39] S. G. Mikhlin and S. Prößdorf. *Singular Integral Operators.* Akademie Verlag, Berlin, 1986.

[40] G. Monegato and S. Prößdorf. Uniform convergence estimates for a collocation and discrete collocation method for the generalized airfoil equation. In A. G. Agarval, editor, *Contribution to Numerical Mathematics*, pages 285–299. World Scientific Publishing Company, 1993. See also the errata corrige in the Internal Report no. 14, 1993, Dipartimento di Matematica, Politecnico di Torino.

[41] W. F. Moss. The two-dimensional oscillating airfoil: A new implementation of the Galerkin method. *SIAM J. Numer. Anal.*, 20:391–399, 1983.

[42] N. I. Muskhelishvili. *Singuläre Integralgleichungen.* Akademieverlag, Berlin, 1965.

[43] P. Nevai. Orthogonal polynomials. *Mem. Amer. Math. Soc.*, 213, 1979.

[44] S. Prößdorf and B. Silbermann. *Numerical Analysis for Integral and Related Operator Equations.* Birkhäuser Verlag, Basel, Boston, Berlin, 1991.

[45] S. Roch. *Lokale Theorie des Reduktionsverfahrens für singuläre Integraloperatoren mit Carlemanschen Verschiebungen.* PhD thesis, TU Karl-Marx-Stadt (now Chemnitz), 1987.

[46] S. Roch and B. Silbermann. *Algebras of Convolution Operators and Their Image in the Calkin Algebra.* Report R-MATH-05/90, AdW der DDR, Berlin, 1990.

[47] S. Roch and B. Silbermann. Index calculus for approximation methods and singular value decomposition. *J. Math. Anal. Appl.*, 225:401–426, 1998.

[48] M. Rosenblum and J. Rovnyak. *Hardy Classes and Operator Theory.* Oxford University Press, New York, Clarendon Press, Oxford, 1985.

[49] B. Silbermann. Lokale Theorie des Reduktionsverfahrens für Toeplitzoperatoren. *Math. Nachr.*, 104:137–146, 1981.

[50] B. Silbermann. Lokale Theorie des Reduktionsverfahrens für singuläre Integraloperatoren. *Z. Anal. Anw.*, 1(6):45–56, 1982.

[51] G. Szegö. *Orthogonal Polynomials.* AMS Colloquium Publications Vol. XXIII, AMS, Providence, Rhode Island, 1939 (Reprint 1985).

[52] G. Vainikko. *Funktionalanalysis der Diskretisierungsmethoden.* Teubner, Leipzig, 1976.

[53] U. Weber. *Lokale Theorie eines Reduktionsverfahrens für singuläre Integraloperatoren auf Intervallen.* Master's thesis, TU Chemnitz, 1992.

Erklärung

Ich erkläre hiermit, daß ich die vorliegende Arbeit selbständig und nur unter Benutzung der angegebenen Literatur und Hilfsmittel angefertigt habe.

Chemnitz, den 19. 7. 1999 Uwe Weber

Thesen zur Dissertation

Weighted polynomial approximation methods for Cauchy singular integral equations in the non-periodic case

eingereicht von Dipl.-Math. Uwe Weber

an der Fakultät für Mathematik
der Technischen Universität Chemnitz

1. The thesis is devoted to the investigation of collocation and Galerkin methods based on weighted polynomials for the approximate solution of singular integral equations on $(-1,1)$ of the type

$$(Au)(x) = a(x)u(x) + \frac{b(x)}{\pi i}\int_{-1}^{1}\frac{u(t)}{t-x}\,\mathrm{d}t = f(x), \qquad x \in (-1,1), \tag{1}$$

where u is the unknown function and a, b, f are given, and of quadrature methods for the solution of

$$[(A+K)u](x) = (Au)(x) + \int_{-1}^{1} k(t,x)u(t)\,\mathrm{d}t = f(x), \qquad x \in (-1,1), \tag{2}$$

where $A = aI + bS$ is as in (1) and k is a given function.

A lot of papers, for instance [3], [4], [5], [6], [7], [9, Chapter 9], has been dealing with approximation methods for such equations, where the coefficients a and b satisfy a Hölder condition. In general they require the determination of the parameters of Gaussian quadrature formulas associated to generalized Jacobi weights that are related to the coefficients a and b of the operator A. The approach in the present paper is aimed at reducing the computational complexity in the preprocessing by using approximation methods that are independent of the concrete equation. Thus, it is always possible to choose pure Jacobi weights, for which there exist explicit (and very simple) expressions for the recurrence coefficients. If one uses the Chebyshev weights, even the weights and nodes of the associated quadrature formulas are known. Our method allows piecewise continuous coefficients and is also applicable to systems.

2. Let $\sigma(x) = v^{\alpha,\beta}(x) = (1-x)^{\alpha}(1+x)^{\beta}$, $v = v^{\gamma,\delta}$ be Jacobi weights the exponents of which are in $(-1,1)$. We will consider Equations (1), (2) in the space $\mathbf{L}^2_{\sigma}$, equipped with the inner product

$$\langle u,v\rangle_{\sigma} := \int_{-1}^{1} u(x)\overline{v(x)}\sigma(x)\,\mathrm{d}x.$$

Denote by p_n^v the orthonormal polynomial of degree n (with positive leading coefficient) with respect to the inner product $\langle .,.\rangle_v$ and let $w_{v,\sigma^{-1}} := \sqrt{\sigma^{-1}v}$. Then, the functions

$$\widetilde{u}_n := w_{v,\sigma^{-1}}p_n^v, \qquad n = 0,1,2,\ldots, \tag{3}$$

form an orthonormal basis in $\mathbf{L}^2_\sigma$. For the approximation methods we want to apply to (1) and (2), the functions (3) will be used as ansatz functions. We need two sequences of projections with respect to the system $\{\widetilde{u}_n\}_{n=0}^\infty$: the Fourier projections P_n^σ given by

$$P_n^\sigma u := \sum_{k=0}^{n-1} \langle u, \widetilde{u}_k \rangle_\sigma \widetilde{u}_k.$$

and the weighted interpolation operators

$$\widetilde{L_n^\sigma} = w_{v,\sigma^{-1}} L_n^v (w_{v,\sigma^{-1}})^{-1} I,$$

where L_n^v is the usual (polynomial) Lagrangian interpolation operator with respect to the nodes x_{jn}^v, which are the zeros of p_n^v. We have $\widetilde{L_n^\sigma} f \to f$ in $\mathbf{L}^2_\sigma$ if f is Riemann integrable and $|f(x)| \leq \text{const}\,(1-x)^{-(1+\alpha)/2+\varepsilon}(1+x)^{-(1+\beta)/2+\varepsilon}$ for some $\varepsilon > 0$. In all our approximation methods we search for an approximate solution $u_n \in X_n := \operatorname{span}\{\widetilde{u}_k\}_{k=0}^{n-1}$. The Galerkin method is described by the approximate equations

$$A_{n,P} u_n := P_n^\sigma A P_n^\sigma u_n = P_n^\sigma f, \tag{4}$$

the collocation method by

$$A_{n,L} u_n := \widetilde{L_n^\sigma} A P_n^\sigma u_n = \widetilde{L_n^\sigma} f. \tag{5}$$

In view of the aim of the practical implementation on the computer, our main emphasis will be on the collocation method.

3. Our main concern is the stability of the sequences $\{A_{n,L}\}$ and $\{A_{n,P}\}$, which will be derived using Banach algebra methods. One of the main tools is a so-called lifting theorem that we give in a version adapted to our concrete situation here. First we introduce the operators

$$W_n^\sigma u := \sum_{k=0}^{n-1} \langle u, \widetilde{u}_{n-1-k} \rangle_\sigma \widetilde{u}_k.$$

Theorem 1 ([11], Satz 3) *Let $\mathcal{A}$ be the C^*-algebra of all sequences $\{A_n P_n^\sigma\}$, $A_n \in \mathcal{L}(X_n)$, for which $A_n P_n^\sigma$, $(A_n P_n^\sigma)^*$, $\widetilde{A_n} := W_n^\sigma A_n W_n^\sigma$ and $\widetilde{A_n}^*$ are strongly convergent, equipped with the supremum norm and component-wise operations. Let $\mathcal{I} \subset \mathcal{A}$ denote the closed ideal of all sequences for which $A_n = P_n^\sigma K_1 P_n^\sigma + W_n^\sigma K_2 W_n^\sigma + C_n$, where K_1, K_2 are compact operators and $\|C_n\| \to 0$ $(n \to \infty)$. Then a sequence $\{A_n\} \in \mathcal{A}$ is stable if and only if the strong limits $A = \text{s-}\lim A_n$ and $\widetilde{A} = \text{s-}\lim \widetilde{A_n}$ are invertible and the coset $\{A_n\} + \mathcal{I}$ is invertible in the quotient algebra $\mathcal{A}/\mathcal{I}$*

4. We show the stability of $\{A_n\} := \{\widetilde{L_n^\sigma}(aI + bS)P_n^\sigma\}$ using Theorem 1. First we verify that this sequence is in $\mathcal{A}$.

Theorem 2 *Let a, b be Riemann integrable and assume that the exponents of the Jacobi weights σ, v satisfy $\alpha \neq \gamma$, $\beta \neq \delta$. Then A_n converges strongly to $A = aI + bS$. If moreover $\gamma > \max\{\alpha - \frac{1}{2}, 0\}$, $\delta > \max\{\beta - \frac{1}{2}, 0\}$, we also have the strong convergence of A_n^*.*

In the special cases $\gamma = \alpha$ or $\beta = \delta$ we have to put additional conditions on the behaviour of b in ± 1 (see Section 4.6).

To be able to investigate the strong convergence of $\widetilde{A_n}$, we restrict ourselves from now on to the situation $v = \varphi := v^{\frac{1}{2},\frac{1}{2}}$.

Theorem 3 *Let a be Riemann integrable and b piecewise continuous with $b(\pm 1) = 0$. Then $\widetilde{A_n}, \widetilde{A_n}^*$ are strongly convergent, and $\widetilde{A} = aI - bw_{v,\sigma}^{-1}Sw_{v,\sigma}I$. In the special case $\sigma = \varphi^{-1}$, the condition $b(\pm 1) = 0$ can be omitted.*

Note that the invertibility of A together with $b(\pm 1) = 0$ implies the invertibility of $\widetilde{A}$. The invertibility of the coset $\{A_n\} + \mathcal{I}$ can be investigated with the local principle of Gohberg and Krupnik. For this end, we make use of a transformation to the unit circle that allows us to apply some results from [8] concerning the collocation method for singular integral equations on the unit circle.

Theorem 4 *Let a, b be piecewise continuous, $b(\pm 1) = 0$. Then $\{A_n\}$ is stable if and only if $A = aI + bS$ is invertible. In the case $\sigma = \varphi^{-1}$ one can dispense with the condition $b(\pm 1) = 0$ and has to require explicitly that both A and $\widetilde{A} = aI - bS$ are invertible.*

5. The collocation method for systems (that is, equations with matrix valued coefficients $\underline{a}, \underline{b}$) is considered only under the condition $\underline{b}(\pm 1) = 0$. In this situation, the stability of the sequence of approximate operators is equivalent to the invertibility of $A = \underline{a}I + \underline{b}S$ and $\widetilde{A} = \underline{a}I - \underline{b}w_{v,\sigma}^{-1}Sw_{v,\sigma}I$.

6. We derive error estimates in a scale of Sobolev spaces $\widetilde{\mathbf{L}}^2_{\sigma,s}$, $s \geq 0$, defined by (compare the spaces $\mathbf{L}^2_{v,s} = \{u \in \mathbf{L}^2_v : \|u\|_{v,s} := \left(\sum_{n=0}^{\infty}(1+n)^{2s}|\langle u, p_n^v\rangle_v|^2\right)^{1/2} < \infty\}$ defined in [2])

$$\widetilde{\mathbf{L}}^2_{\sigma,s} = \left\{ u \in \mathbf{L}^2_\sigma : \|u\|_{s,\sim} := \left(\sum_{n=0}^{\infty} (1+n)^{2s} |\langle u, \widetilde{u}_n \rangle_\sigma|^2 \right)^{1/2} < \infty \right\}.$$

One can show that for $s > \frac{1}{2}$ these spaces are continuously embedded in spaces of weighted continuous functions, which makes it possible to obtain weighted uniform convergence results.

Theorem 5 *Assume that the solution u of (1) belongs to $\widetilde{\mathbf{L}}^2_{\sigma,s}$, $s > \frac{1}{2}$, and $A \in \mathcal{L}(\widetilde{\mathbf{L}}^2_{\sigma,r})$ for some r, $\frac{1}{2} < r \leq s$. If the sequence $\{A_{n,L}\}$ is stable, we have the estimation*

$$\|u_n - u\|_{t,\sim} \leq \text{const}\, n^{t-s} \|u\|_{s,\sim}, \quad 0 \leq t \leq s.$$

where $u_n \in \operatorname{im} P_n^\sigma$ is the solution of (5).

In the thesis (Section 7) some conditions are given under which $aI + bS$ is bounded in $\widetilde{\mathbf{L}}^2_{\sigma,s}$. Here we give only the result for the case $\sigma = \varphi^{-1}$.

Theorem 6 *Let s be an integer and assume that $\varphi^k a^{(k)}, \varphi^{k+1}(\varphi^{-1}b)^{(k)} \in \mathbf{L}^\infty$, $k = 0, \ldots, s$. Then A is bounded on $\widetilde{\mathbf{L}}^2_{\sigma,s}$. If moreover A is invertible in $\mathbf{L}^2_\sigma$, $a, b \in \mathbf{C}^{s,\eta}$ for some $\eta > 0$ and $b^{(k)}(\pm 1) = 0$, $k = 0, \ldots, s$, then A^{-1} is also bounded on $\widetilde{\mathbf{L}}^2_{\sigma,s}$.*

7. We show that the implementation of the collocation method leads in a wide class of cases to the solution of a system of linear equations with a Löwner-like matrix. This allows to apply a fast algorithm with a complexity of $O(n^2)$ operations.

8. In (2) we assume that $w_{v,\sigma}^{-1}(t)\,k(t,x) \in \mathbf{C}([-1,1]^2)$. Equation (2) is solved approximately by a quadrature method of the form

$$A_{n,K}u_n := \widetilde{L_n^\sigma}(A+K_n)P_n^\sigma u_n = \widetilde{L_n^\sigma} f, \tag{6}$$

where

$$(K_n u)(x) = \int_{-1}^{1} [w_{v,\sigma} L_n^v w_{v,\sigma}^{-1} k(\cdot,x)](t)u(t)\,dt.$$

Theorem 7 *The sequence $\{\widetilde{L_n^\sigma}(A+K_n)P_n^\sigma\}$ is stable if and only if the sequence $\{\widetilde{L_n^\sigma} A P_n^\sigma\}$ is stable and the operator $A+K$ is invertible in $\mathbf{L}_\sigma^2$.*

Theorem 8 *Assume that $\{A_{n,K}\}$ is stable, $A \in \mathcal{L}(\widetilde{\mathbf{L}}_{\sigma,r}^2)$ for some $r > \frac{1}{2}$. Furthermore, let $w_{v,\sigma}^{-1} k(\cdot,x) \in \mathbf{L}_{v,s}^2$ uniformly with respect to $x \in [-1,1]$, $N_s \in \mathbf{L}^1(-1,1)$, where $N_s(t) = \sigma^{-1}(t)\,\|k(t,.)\|_{s,\sim}^2$, and $s > \frac{1}{2}$. Then the following error estimate of the quadrature method (6) is valid if $u \in \widetilde{\mathbf{L}}_{\sigma,s}^2$:*

$$\|u_n - u\|_{t,\sim} \le \text{const}\, n^{t-s}\, \|u\|_{s,\sim}$$

for $0 \le t \le s$.

9. The structure of the system matrix mentioned in 7 is destroyed by the perturbation occurring in the quadrature method for (2). We consider two fast algorithms that allow us nevertheless to maintain a computational complexity of $O(n^2)$, together with an error estimate like in Theorem 8, using the fast solution of the pure collocation equations and some mapping properties of the operators involved in Sobolev spaces.

10. The Amosov-type algorithm ([1]) works as follows: Choose some integer $m < n$, let $Q_m^\sigma = I - P_m^\sigma$, and determine the approximate solution $u_n \in X_n$ as follows:

    1. **Collocation method for unperturbed equation:** Put $Q_m^\sigma u_n = Q_m^\sigma v_n$, where $A_n v_n = \widetilde{L_n^\sigma} f$.

    2. **Correction step:** Put $P_m^\sigma u_n = w_m$, where $(A_m + \widetilde{L_m^\sigma} K_m)w_m = \widetilde{L_m^\sigma} f - \widetilde{L_m^\sigma} A Q_m^\sigma v_n$.

Theorem 9 *Let $\frac{1}{2} < t \le s$, $u \in \widetilde{\mathbf{L}}_{\sigma,s}^2$, $\delta \ge \frac{s-t}{2}$, and let the following assumptions be satisfied: Assume that the sequence $\{A_{n,K}\}$ is stable in $\mathbf{L}_\sigma^2$, $A \in \mathcal{L}(\widetilde{\mathbf{L}}_{\sigma,s}^2)$, $A^{-1} \in \mathcal{L}(\widetilde{\mathbf{L}}_{\sigma,s+\delta}^2)$, $N_{s+\delta} \in \mathbf{L}^1$. Furthermore, let $(A+K)^{-1} \in \mathcal{L}(\widetilde{\mathbf{L}}_{\sigma,s}^2)$, and let, finally, $w_{v,\sigma}^{-1} k(\cdot,x) \in \mathbf{L}_{v,s+\delta}^2$ uniformly with respect to x. Then, if we choose the order m of the system in the correction step such that $m \sim n^{2/3}$, we can obtain an error estimate*

$$\|u_n - u\|_{t,\sim} \le \text{const}\, n^{t-s}\, \|u\|_{s,\sim}$$

with a computational complexity of $O(n^2)$ operations.

11. The second algorithm is a two-grid iteration method that works as follows: For fixed positive integers $0 < m < n$ we choose $u_{n,0} \equiv 0$ and put

$$u_{n,j+1} := v_{n,j} + w_{m,n,j}, \qquad j = 0, 1, 2, \ldots,$$

where $w_{m,n,j} \in \operatorname{im} P_m^\sigma$ is determined by

$$\widetilde{L_m^\sigma}(A + K_m) w_{m,n,j} = \widetilde{L_m^\sigma} K_n (u_{n,j} - v_{n,j})$$

and $v_{n,j} \in \operatorname{im} P_n^\sigma$ is the solution of

$$A_n v_{n,j} = \widetilde{L_n^\sigma} f - \widetilde{L_n^\sigma} K_n u_{n,j}.$$

One can rewrite this algorithm in the form

$$u_{n,j+1} = B_n \widetilde{L_n^\sigma} f + C_n u_{n,j}$$

with certain operators B_n, C_n, and under suitable assumptions the estimation $\|C_n\|_{\mathcal{L}(\widetilde{\mathbf{L}}^2_{\sigma,t})} \le \text{const}\, m^{t-s}$ for sufficiently large m and n can be shown. Thus, Banach's fixed point theorem leads to the following result:

Theorem 10 *Assume that $\{A_{n,K}\}$ is stable (in $\mathbf{L}^2_\sigma$), $A \in \mathcal{L}(\widetilde{\mathbf{L}}^2_{\sigma,r})$ for some $r > \frac{1}{2}$, $u \in \widetilde{\mathbf{L}}^2_{\sigma,s}$, $A^{-1}, (A+K)^{-1} \in \mathcal{L}(\widetilde{\mathbf{L}}^2_{\sigma,s})$, $s > \frac{1}{2}$, $N_s \in \mathbf{L}^1$ and $\left\|w_{v,\sigma}^{-1} k(\cdot, x)\right\|_{v,s} \le \text{const}$ for all $x \in (-1,1)$. Then the iteration process converges to the solution u_n of (6), and we have the error estimate*

$$\|u_n - u_{n,j}\|_{t,\sim} \le \text{const}\, m^{j(t-s)} \|u_n - u_{n,0}\|_{t,\sim} \qquad \text{for all} \quad 0 \le t < s.$$

Thus, choosing $m = O(n^{1/2})$ and performing two iteration steps, we obtain an error of the same order as the discretization error according to Theorem 8 with a complexity of $O(n^2)$ operations.

The computational expense is somewhat higher than that in the Amosov algorithm, though of the same asymptotic order. On the other hand, the assumptions on the data are weaker, and the convergence order is maintained also for $0 \le t \le \frac{1}{2}$.

12. To investigate the stability of the sequence $\{A_{n,P}\}$ from (4) in the space $\mathbf{L}^2$ (we have $v = \varphi$, $\sigma \equiv 1$), we make a transformation to the Hardy space $\mathbf{H}^2(\mathbb{T})$, and it turns out that the transformations of the singular integral operators can be expressed in terms of Toeplitz and Hankel operators. This allows to apply the results from [10] concerning the finite section method for singular integral operators with Carleman shift. One gets the following result ($P_\mathbb{T}$ denotes the orthogonal projection of $\mathbf{L}^2(\mathbb{T})$ onto $\mathbf{H}^2(\mathbb{T})$, and $Q_\mathbb{T} = I - P_\mathbb{T}$):

Theorem 11 *For $a, b \in \mathbf{PC}$ and $A = aI + bS$ or $A = aI + SbI$, the operator sequence $\{P_n^\sigma A P_n^\sigma\}$ is stable in $\mathbf{L}^2(-1,1)$ if and only if the following conditions are satisfied:*

(a) A is invertible on $\mathbf{L}^2$,

(b) $[\widehat{a} - \widehat{b}\psi] P_\mathbb{T} + Q_\mathbb{T}$ is invertible on $\mathbf{L}^2(\mathbb{T})$, and

(c) for every $x \in (-1,1)$ the origin lies outside the triangle which is formed by the points $\dfrac{c^-(x)}{c^+(x)}$, $\dfrac{\widetilde{c}^-(x)}{\widetilde{c}^+(x)}$, and 1,

where $c^+(x) := a(x-0) - b(x-0)$, $c^-(x) := a(x+0) - b(x+0)$, $\widetilde{c}^+(x) := a(x-0) + b(x-0)$, $\widetilde{c}^-(x) := a(x+0) + b(x+0)$.

References

[1] B. A. Amosov. On the approximate solution of elliptic pseudodifferential equations on a smooth closed curve (Russian). *Z. Anal. Anw.*, 9:545–563, 1990.

[2] D. Berthold, W. Hoppe, and B. Silbermann. A fast algorithm for solving the generalized airfoil equation. *J. Comp. Appl. Math.*, 43:185–219, 1992.

[3] M. L. Dow and D. Elliott. The numerical solution of singular integral equations over $(-1,1)$. *SIAM J. Numer. Anal.*, 16(1):115–134, 1979.

[4] D. Elliott. The classical collocation method for singular integral equations. *SIAM J. Numer. Anal.*, 19(4):816–832, 1982.

[5] D. Elliott. Orthogonal polynomials associated with singular integral equations having a Cauchy kernel. *SIAM J. Math. Anal*, 13(6):1041–1052, 1982.

[6] D. Elliott. A comprehensive approach to the approximate solution of singular integral equations over the arc $(-1,1)$. *Journal of Integral Equations and Applications*, 2:59–94, 1989.

[7] P. Junghanns and B. Silbermann. Theorie der Näherungsverfahren für singuläre Integralgleichungen auf Intervallen. *Math. Nachr.*, 103:199–244, 1981.

[8] P. Junghanns and B. Silbermann. Local theory of the collocation method for singular integral equations. *Integral Equations Operator Theory*, 7:791–807, 1984.

[9] S. Prößdorf and B. Silbermann. *Numerical Analysis for Integral and Related Operator Equations.* Birkhäuser Verlag, Basel, Boston, Berlin, 1991.

[10] S. Roch. *Lokale Theorie des Reduktionsverfahrens für singuläre Integraloperatoren mit Carlemanschen Verschiebungen.* PhD thesis, TU Karl-Marx-Stadt (now Chemnitz), 1987.

[11] B. Silbermann. Lokale Theorie des Reduktionsverfahrens für Toeplitzoperatoren. *Math. Nachr.*, 104:137–146, 1981.

Lebenslauf

Uwe Weber

geboren am 15. Juni 1968 in Karl-Marx-Stadt (jetzt Chemnitz) als Sohn des Behördenangestellten Wolfgang Weber und der Konfektionärin Margitta Weber geb. Friedel.

Wohnort	Friedrich-Hähnel-Str. 70, 09120 Chemnitz
Schulbildung	
Sept. 1975–Aug. 1985	Polytechnische Oberschule in Karl-Marx-Stadt
Sept. 1985–März 1986	ABF Halle
April 1986–Aug. 1987	Erweiterte Oberschule in Karl-Marx-Stadt
Studium	
Sept. 1987–Sept. 1992	Diplomstudium Mathematik an der TU Karl-Marx-Stadt/Chemnitz, Spezialisierungsrichtung: Numerische Analysis
Okt. 1992–Dez. 1992	Stipendium der Studienstiftung des deutschen Volkes
Beruflicher Werdegang	
Jan. 1993–Sept. 1999	Wissenschaftlicher Mitarbeiter an der Fakultät für Mathematik der TU Chemnitz
seit Okt. 1999	Wissenschaftlicher Mitarbeiter an der Fakultät für Mathematik und Informatik der TU Bergakademie Freiberg